Living with Our Sun's Ultraviolet Rays

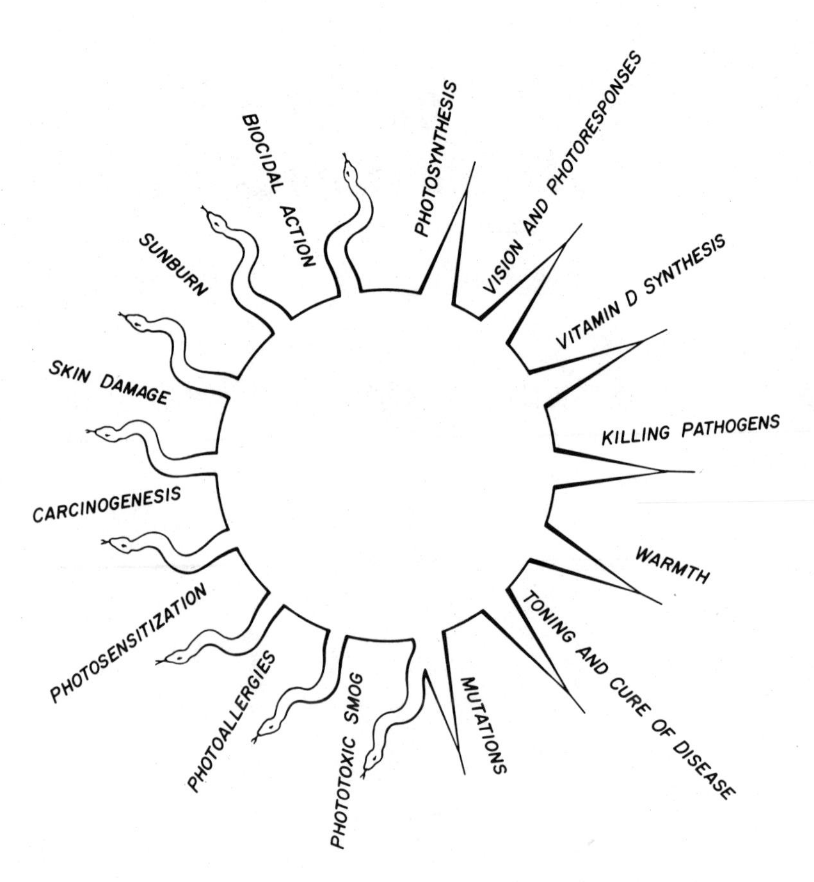

Our sun has two faces–one good and one bad, like Janus, the Roman god who guards the gateway to heaven

Living with Our Sun's Ultraviolet Rays

Arthur C. Giese

Professor of Biology, Emeritus
Stanford University

PLENUM PRESS • NEW YORK AND LONDON

Library of Congress Cataloging in Publication Data

Giese, Arthur Charles, 1904-
Living with our sun's ultraviolet rays.

Bibliograhy: p.
Includes index.
1. Ultra-violet rays–Physiological effect. 2. Skin, Effect of radiation on the. I. Title. [DNLM: 1. Ultraviolet rays. 2. Radiobiology. WN610 G455L]
QP82.2.U4G53 613.1'9 75-44225
ISBN 0-306-30883-5

A Division of Plenum Publishing Corporation
227 West 17th Street, New York, N.Y. 10011

United Kingdom edition published by Plenum Press, London
A Division of Plenum Publishing Company, Ltd.
Davis House (4th Floor), 8 Scrubs Lane, Harlesden, London, NW10 6SE, England

Printed in the United States of America

To

Philip A. Leighton

who stimulated my interest in the chemical and biological effects of light, and to the memory of

Warder C. Allee

who aroused my curiosity in environmental biology

Preface

Sunlight is part of everyday life and we accept it as good—and good it is in a number of ways. The sun is our source of warmth, and of the light by which we see. It is, in fact, the source of the energy with which life continues on earth. It furnishes energy for photosynthesis, and the products of photosynthesis constitute our food, building materials, and fuel.

A steady state of balance and fine interrelationships exists between life on earth and all the forces and stresses in nature. This book will pinpoint the balance and relationships we share with sunlight. Our primary focus will be on the ultraviolet radiation of the sun, and on the ultraviolet photobiology of life on earth. This is the story of the effects of the sun's ultraviolet radiation, both good and bad, on all of us and all of life. We will explore the nature of the sun's ultraviolet radiation as it reaches the earth's surface today, and as it probably affected the earth in the distant past; and examine the effect of such radiation on all life, unicellular organisms as well as multicellular plants and animals. The effects of the sun's ultraviolet rays are primarily a result of their action upon cells, and secondarily, a result of their interactions between cells. The cell of a multicellular organism—man included—is also part of the tissue of an organ, and the organ is part of the whole organism. Thus, an effect on cells also leads to effects on the entire organism. Therefore, the major emphasis in this account and, I believe, the key to the understanding of our subject, is the cell's response to the sun's ultraviolet radiation.

The earth's population is increasing steadily and the staggering problems of pollution continue to mount—pollution of the air we breathe, the water we drink, the food we eat, the total environment of all living things. Dramatic events in our recent past have brought these dilemmas to the public's attention. But life on earth must face still another crisis brought on by pollution of another sort, a kind of pollution not easily controlled, but hopefully, not yet beyond control. This is the pollution of our high atmosphere, the stratosphere, by a proposed large fleet of commercial supersonic transports (SSTs), by gases used in refrigerators and as propellants in spray cans, and by atomic bomb tests, all of which may profoundly change the quantity and even the quality of the ultraviolet radiation that reaches the earth.

It may seem to those unacquainted with all the facts that we need have little concern about the stratosphere. But it is in this very stratosphere that a layer of ozone removes from incoming sunlight the photochemically powerful ultraviolet rays so devastating to life, and transmits only that part of the ultraviolet spectrum to which, in one way or another, life has accommodated itself.

This book is intended for the intelligent reader who has little knowledge of photobiology but who has some interest in science, and so avoids many of the controversies in the scientific literature. For that reason documentation is minimal and references, mainly to recent literature, are given at the end of each chapter.

Although the book is integrated as a whole, each chapter is organized as a unit for greater simplicity. If Chapters III and V to VIII are dominated by information on the effects of sunlight's ultraviolet radiation on man, it is not by choice. I have searched without much success for information on the effects of such radiation on plants and animals. Those photobiologists who in the past were interested in the effects of ultraviolet radiation on organisms have generally used lamps that included much shorter wavelengths of ultraviolet radiation than those present at the earth's surface. Their findings, therefore, have little bearing on the topic of major interest here.

If it seems that some items of interest have been omitted or misinterpreted, it was not intended. Probably no one person, at least not I, can fully grasp the vast literature our theme has engendered.

I am indebted to many individuals for discussions on some of the topics covered, especially to my colleagues, Alvin Cox, Farrington Daniels, Jr., Philip Hanawalt, John Lee, Eleanor McDonald, David Regnery, Kendric Smith, Steven H. Tomson, and Frederick Urbach. I am grateful also for suggestions from readers of the manuscript, Alan Bruce, Howard Ducoff, Carl May, William Odum, and Kendric Smith;

and for continuous encouragement from Kendric Smith. I am equally indebted to Gretchen Montalbano and Arlene Novak for their expert typing of a difficult manuscript, to Teppy Williams for artwork, and to Sundance Acacia for many of the graphs. Thanks are also due to many authors and publishers for permission to use adaptations of their figures.

Most of all, I am indebted to my wife, Raina, for her forbearance and telling criticisms, and to Robert Ubell of Plenum Publishing Corporation for his critical editing of the final manuscript.

Arthur C. Giese

Stanford University
February, 1976

Contents

1

The Sun, Sun Myths, and Sun Worship

Most of us take the sun for granted. We know it to be a heavenly body subject to the same laws of physics that govern events on earth. For example, we know that the earth's axis is not vertical to the earth's orbit around the sun; that the resultant oblique angles at which the sun's rays strike the earth cause seasonal variation in radiation (as Figure 1:1 shows); and that the quantity is greatest when the days are longest. We have confidence that the sun will rise each morning and set each evening. We know, as Figure 1:2 demonstrates, that the length of the day will vary regularly; that in the Northern Hemisphere it is maximal at the summer solstice (June 21) and minimal at the winter solstice (December 22); and that the length of day is equal to the length of night at the vernal equinox (March 21) and the autumnal equinox (September 22). The sun presents us with our major cues to time and season, and figures prominently in agricultural practices as well.

We can predict that every 18 years and 11⅓ days the moon, in its orbit around the earth, will be positioned directly between the sun and the earth and will eclipse the sun and create darkeness over a limited area of the earth, even during the brightest part of the day. We do not fear solar eclipses, as many primitive peoples did, because we know that the earth will soon be out of the shade of the moon. A total eclipse of the sun can be viewed only briefly on earth, and as we see in Figure 1:3, it can be viewed only over a circular area with about a 30-mile diameter because of the small diameter of the moon compared to that of the sun. For that reason, astronomers are willing to travel long distances to the right spots

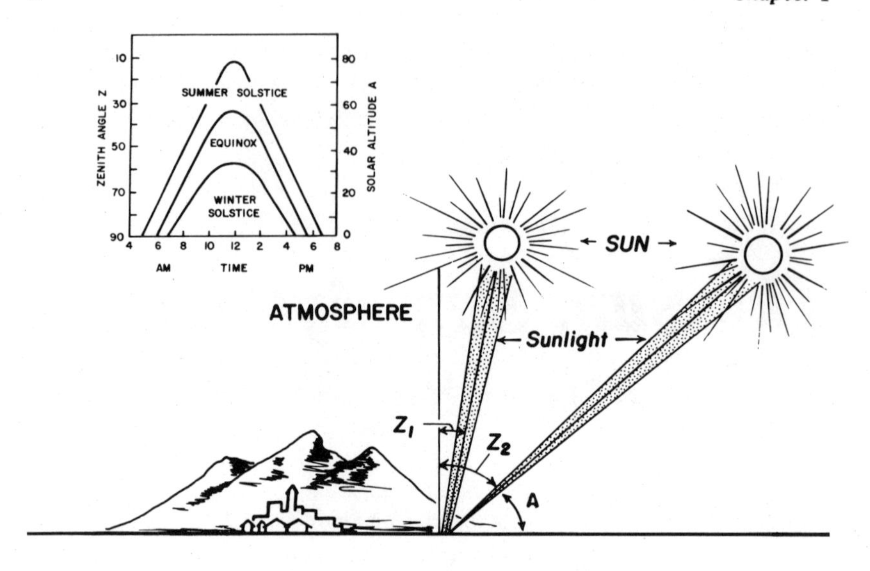

FIGURE 1:1. Diagram showing the effect of season, latitude, and time of day on the depth of atmosphere through which sunlight must pass before reaching a point on the surface of the earth. At noon at the summer solstice, the sun has the smallest zenith angle (Z_1), and the angle at noon increases to a maximum at the winter solstice (Z_2), the angle between these two extremes being characteristic of the autumn and spring equinoxes. At any time of the year the sun's angle increases with the latitude (as one moves toward the poles from the equator) and during the day on either side of high noon, four hours being equivalent to about 60° for the angle Z. The thicker the atmosphere through which sunlight must pass, the lower its intensity at the surface of the earth; thus, any increase in the angle Z results in a lower intensity. Version by Carl May. Insert shows the relationship between the solar zenith angle Z and the time of day at Los Angeles. The solar altitude A is the angle the sun makes with the horizontal ($A = 90° - Z$). After Leighton, *Photochemistry of Air Pollution,* Academic Press, New York, 1961, p. 10.

for viewing eclipses. Much of what astronomers know about the sun's atmosphere and corona, the aura of light seen around the periphery of the eclipsed sun, comes from such limited studies. This corona has a temperature higher than the surface of the sun for reasons not yet fully understood.

Sun and Sunlight

What detailed knowledge we have of the sun and of our universe is relatively recent, for until the middle of the sixteenth century it was still thought that the earth was the center of our universe and that the sun revolved around the earth. But, in 1543, when the Polish-born astronomer Nicolaus Copernicus finally succeeded in publishing his treatise *De Revolutionibus Orbium Coelestium,* he clearly showed that many observations on the positions of the planets with respect to the sun and the earth that were difficult to explain by means of the geocentric theory—that the earth is the center of the universe—could be explained simply by the revolutionary new heliocentric theory. Copernicus proposed that the sun is the center of our planetary system. He suggested that the earth rotates on its axis and, like the other planets of our solar system, revolves about the sun. The heliocentric view of the solar system stimulated the growth of astronomy and furthered our understanding of the sun.

We now know that the sun is indeed the center of our planetary system and that it is about 93 million miles from the earth. It is only one

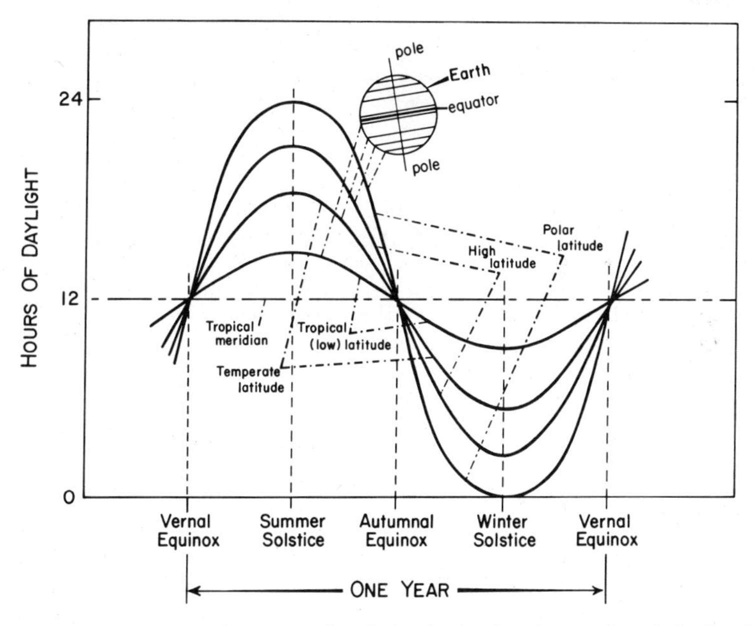

FIGURE 1:2. Curves showing seasonal variation in day length at various latitudes. Low latitudes are those (north and south) nearest the equator with latitudes becoming progressively "higher" toward the poles. Note that during a year the extremes of light and darkness occur at the higher latitudes, whereas the variation is slight at lower latitudes. One year is the period between two spring equinoxes. The fall equinox is shown midway between the two spring equinoxes. Day length is constant at the tropical meridian. Version by Carl May.

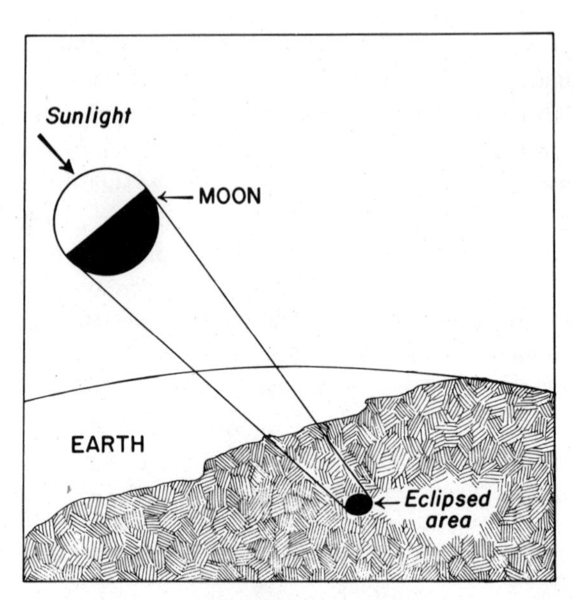

FIGURE 1:3. Eclipse of the sun by the moon interposed between sun and earth. After Menzel, *Our Sun,* Harvard University Press, Cambridge, Massachusetts, 1959, p. 225.

of the many suns in our galaxy, and our galaxy is only one of some 260 million galaxies in the universe. Our sun, a yellow star, with a diameter 109 times that of the earth and a mass about 300,000 times greater than the earth, is neither the largest nor the hottest star in our galaxy. A blue star is about a hundred times as bright as our sun, and a red star only about one-hundredth as bright as our sun. The temperature of the sun's surface is about 5600°K (absolute), and the temperature of the core is many millions of degrees absolute. The white-hot filament of an ordinary household tungsten light bulb is about one-third the surface temperature of the sun.

Within the sun, thermonuclear reactions transmute about 564 million tons of hydrogen into 560 million tons of helium every second with the conversion of 4 million tons of matter into radiant energy. Sunlight provides the surface of the earth's atmosphere with almost two calories per square centimeter per minute. Although a third of this is absorbed in the atmosphere, on a worldwide basis this amount of energy is more than sufficient to supply us with all the heat and power we need when we learn how to harness and store it effectively for our use.

Only a very small fraction of the photic (or light energy) in sunlight striking the earth is stored by green plants as chemical energy. We use plants as food and store them, and especially their seeds, to supply our needs during seasons unfavorable for plant growth. We use the chemical energy stored in plants in various industries and as fuel. We also use the "fossil" energy stored by plants of ancient geological eras in coal, oil, and gas. Our other sources of energy—water power, wind, and tidal energy—are ultimately derived from the sun; and atomic energy, liberated from elements in the crust of the earth, is ultimately derived from the sun, too. This is because the earth was probably formed from the original solar nebular cloud of matter that gave rise to the solar planetary system.

The radiation absorbed daily from the sun by land and sea, except that stored by plants, must be re-emitted, since the overall temperature of the earth does not change; such radiation from the earth is practically all re-emitted as heat rays in the infrared part of the radiation spectrum. The sun, then, is ultimately the source of virtually all the earth's energy, a circumstance that is now drawing increasing attention from those concerned with man's energy supply.

Astronomers predict that the sun will long continue to supply the earth with light and heat at the present rate, being only halfway through its recently estimated lifetime of 10 billion years. Its core will then contract as the sun becomes a red giant star and its outer layers will puff past the orbit of Venus, heating the earth as they do so.

When sunlight is examined with a spectroscope—an instrument in which the component wavelengths are dispersed by prisms (much as sunlight is dispersed by water droplets acting as prisms in the formation of a rainbow)—dark lines appear throughout the spectrum. These are called Fraunhofer lines, after the investigator who first observed and measured them. They are the absorption lines of various elements in the sun's atmosphere, elements that "soak up" wavelengths of the continuous spectrum coming from the sun's outer surface, the photosphere. Fraunhofer lines identify the elements in the sun's atmosphere, and show that practically all the elements present on earth are found in the sun. Spectra of the few missing elements are probably masked in some way. Thus, the same chemical elements are present in the earth's crust, in meteorites, in the atmosphere of the sun, and in the atmosphere of other stars (although not in the same proportions). Even cosmic ray particles, which rain on us from outer space, appear to be atomic nuclei of hydrogen and other elements present on earth that have had their electrons stripped away.

When the surface of the sun is examined through a high-powered telescope, it shows sunspots that appear to increase to a maximum of eleven-year periods, as indicated in Figure 1:4. A sunspot has a dark center surrounded by a bright border; this center has a temperature about 1500° lower than the sun's surface. Sunspots are the result of magnetic fields that periodically appear in the sun and can often be detected preceding development of the sunspot. Solar flares of much higher tempera-

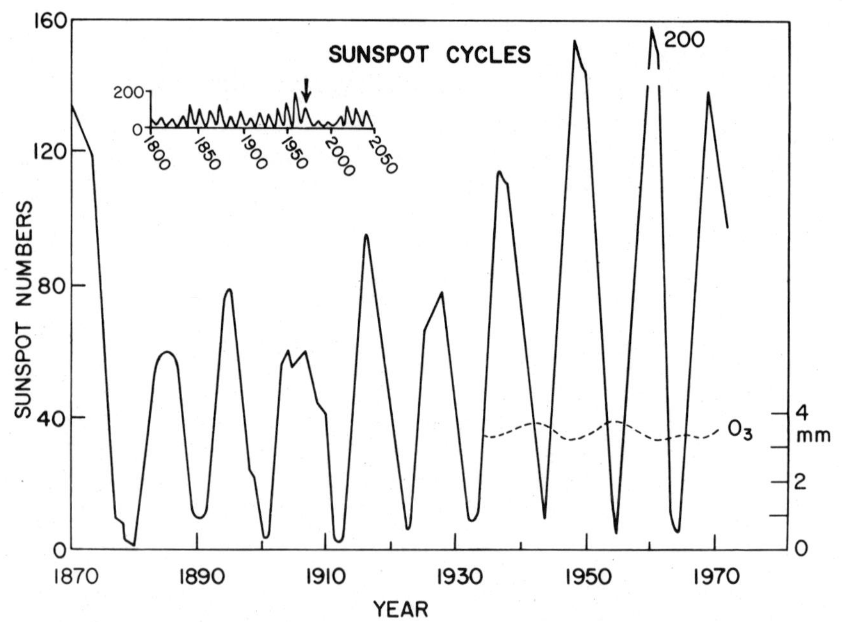

FIGURE 1:4. Variation in sunspot numbers with time. Note that the peaks come at approximately eleven-year intervals, and that the peak number is not always the same. Data from Gibson, *The Quiet Sun,* NASA, 1973, p. 34. The ozone data to the right of the graph are averages for the earth's atmosphere and vary between 3 and 3.5 mm partial pressure ("thickness") out of the 760 mm atmospheric pressure. Note that the ozone peak expected about 1963 did not appear, presumably because of the extensive atomic bomb tests carried out by the United States and the Soviet Union (see Chapter 9). The insert, a condensed version of such cycles for over two hundred years, shows both the cycles recorded, and beyond the arrow, those predicted for the future. Data from Cohen and Lintz, *Nature* 250 (1974), p. 398.

ture than the sun's surface and reaching up to thirty thousand miles from the surface of the sun accompany sunspot activity. These, too, are probably a result of the electrical activity accompanying magnetic fields and also have an eleven-year cycle. They emit intense ultraviolet radiation, x-rays, and cosmic rays, including atomic particles that reach the earth's atmosphere and disturb both the earth's magnetic field and the ionized layers in our atmosphere that reflect radio waves. In addition, solar flares emit intense radio waves that interfere with radio-wave transmission on earth. Although in local areas of the sun's surface the radiation may vary in intensity during sunspot activity, the amount of radiation emitted from the whole sun varies less than 1 percent (possibly as little as 0.2 percent) over a whole year.

Differences in day length and the obliquity of the sun's rays, as well as the differential heating of the oceans and the large-scale movements of sea water and air masses, are responsible for the changes in weather on the earth. It is thought that the changes in weather over a period of years are associated, possibly causally, with the sunspot cycle; however, the basis of these relations is not well understood.

Sun Myths and Worship

Ancient peoples knew intuitively the importance to life of the sun's heat and light; they needed its heat for warmth and its light to see. After the advent of agriculture, heat and light were needed to produce crops on which society depended. But apparently ancient cultures did not take the sun for granted; they made daily observations and recognized the sun's daily and seasonal regularity. in the records of cultures with written traditions, we find that even eclipses were predicted, but neither the laws of astrophysics that govern the sun's movement nor the probable physical nature of the sun and moon were known to the ancients. The sun was conceived as either an object transported by man or animals, or as a god with characteristics much like man; and to ensure its daily presence, the sun or the beings controlling it had to be kept in a favorable mood.

Primitive peoples believed that to keep in the sun's favor they had to propitiate it to make sure that it would rise and set regularly and provide a warm season to grow their crops. In some societies propitiation was ceremonial and token in nature, in others it called for the sacrifice of animals, and in still others it demanded the highest sacrifice of all—that of fellow human beings.

When nature cooled off in winter, Pueblo Indians made ceremonial fires, symbolically providing the sun with heat. To keep up the sun's strength for its daily journey, the Aztecs supplied it with the hearts of sacrificial victims. Eskimos and ancient Egyptians, in legend, rowed the sun at night from the west back to the east to make sure it would have the strength to rise again the next morning. In Greek, Roman, and other mythology, the sun god drove a horse-drawn chariot across the skies. Legends and practices of sun propitiation were prominent even in societies with well-developed astronomy, as in the ancient Egyptian, Mayan-Aztec, and Incan civilizations.

More primitive cultures feared eclipses of the sun and looked upon them as disasters to life and crops. Eclipses were explained in many ways, among them legends that tell of demons, vicious beasts, or people trapping the sun. The eclipsed sun was redeemed by totem animals, heroes, or gods. Sometimes, thanksgiving ceremonies accompanied the sun's "rescue."

In mythology, fire is often a relative of the sun and sometimes depicted as the son of the sun. Fire was revered as a relief from darkness, but people recognized its weakness compared to the glory of the sun's light. Fire was considered a gift of the gods, and legends and ceremonies about fire are extensive and reach into our own times. Fire is still central in the worship of the Parsees of India, descendants of the Persian worshippers of Mithras. Witness the Olympic Games ceremony of the torch, the use of candles in religious observances or by fraternities and sororities, and the lighting of matches at athletic games, rock concerts, and other events.

So pervasive were sun legends that they entered into almost every culture. Northern Europe was no exception, but the details of sun worship in that part of the world are lost to us, for their oral traditions came to an end after the advent of Christianity.

Sun mythology profoundly influenced decorative art throughout the world. The nimbus (halo), with rays or spikes radiating from the head to represent the sun's rays, was widely used in the ancient civilized world for heroes or rulers who were identified with or traced their lineage from the sun or sun gods. Even Louis XIV of France called himself the Sun King. Solar disks, sometimes winged, represented the sun god and were often made of gold. Because of its color and nontarnishing quality, gold became the noble metal. The multipointed golden crowns of monarchs carried solar symbolism, and the golden "solar" robes they often wore also showed the sun's glory reflected on royalty. According to some historians, the cross, which was in use long before Christianity, originally symbolized the sun and its rays. A circle symbolizes the sun at the center of the Celtic cross.

In architecture, too, solar myths played a crucial role, especially where ceremonies honoring the sun were performed. The trilithons and circles of stones at Stonehenge and at the other eighty henges in England and Brittany, the pyramids, and the sun temples in Egypt, Central America, and the Andes were built according to reference points of a compass. Figure 1:5 indicates how many of these ceremonial areas faced east to facilitate worship of the rising sun. These structures were also oriented to provide keener astronomical observations, which were generally made by the high priests.

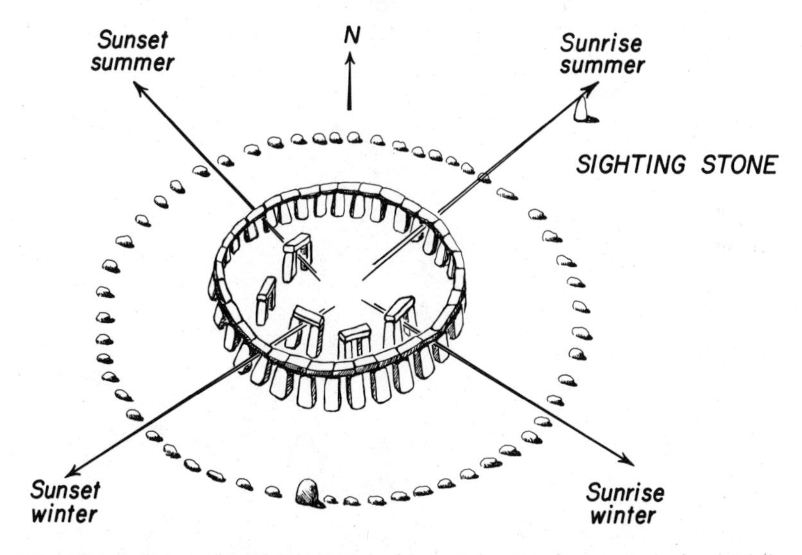

FIGURE 1:5. Orientation of Stonehenge with regard to the summer and winter solstices. The viewer could sight through five archways (trilithons, the triple stones forming an arch) inside the circle both summer and winter solstice positions of the rising and setting sun and moon. The outer ring of fifty-six stones may have been an eclipse computer, according to Gerald Hawkins of the Smithsonian Astrophysical Observatory. After Edson, *Worlds Around the Sun,* American Heritage Publishing, New York, 1969, p. 7.

Sun worship and solar ceremonies were widespread in the ancient world, and usually took place at the time of the solstices and equinoxes. The sun god of the early cultures was represented in a number of ways, each symbolic of a different aspect of the sun—heat, light, or power; and, as in ancient Greece, at different times of day and different seasons. Often, a multiplicity of sun gods arose from the coalescence of many tribes previously separated from one another, each with its own favorite god. Sometimes, as in Egypt, a single supreme god emerged, Amen-Ra (or Ra).

Sun worship may have reached its height in Egypt where, during the reign of Amenhotep IV, who was also called Amenophis IV (1375?–1357 B.C.), the sun god became the only one recognized by the state, thus replacing a complex and elaborate pantheon of gods. The relief in Figure 1:6 shows Amenhotep IV holding the sun's rays in his hands. The sun may have appealed to the Egyptians particularly because of its daily "renewal," and because it provided implications of resurrection and immortality. The Pharaohs identified themselves with the sun god, and attached Ra to their names and later even considered themselves Ra incarnate. They built obelisks symbolizing the sun's rays and constructed sun temples, the most famous and magnificent of which is the one at Karnak. The temples are carefully oriented to permit entry of sunlight through the gateway at the time of the winter solstice.

Helios, the son of a Titan, was the sun god of the ancient Greeks. Homer describes him driving a golden chariot across the heavens. About 600 B.C., he was depicted with an overhead disk; later, spiked rays were added to the representation. Helios is shown on Rhodes coins dating back to before 304 B.C. with streaming locks; and in later artifacts, he wears a spiked halo. Helios was probably the most important god in the pantheon, and it is believed that the Colossus of Rhodes in the island's harbor represented him.

Phoebus Apollo, son of Zeus and the god of light, healing, music, poetry, prophecy, and manly beauty, represented the sun's rays and was later identified with the sun god Helios. The ancient Greeks worshiped

FIGURE 1:6. The sun, upper right, sends its rays to earth dwellers. The rays end in human hands, two of which bear the Egyptian symbol of life. Note that the hands also extend to the flowering plants. Photograph, courtesy of the Metropolitan Museum of Art.

him as creator of the warmth of the springtime and their protector from danger and disease during unfavorable seasons.

The Romans also show Apollo in a chariot. The charioteer is a constant motif in the art of the ancient world, even in that of India, Turkestan, and China. A bronze and gold sun chariot has even been unearthed in a Danish bog near Tröndholm, Zealand.

Persian bas-reliefs show the heads of kings and gods encircled by a rayed halo, symbolic of the power of the sun and Mithras, the ancient Persian god of light and enlightenment. Mithras, who is also shown ascending to heaven in a chariot, was closely associated with the sun.

The Romans referred to Mithras as *sol invictus,* the unconquerable sun. Later, many of the solar aspects of Mithraism, which preceded Christianity in Rome, were absorbed into Christian symbolism. An early mosaic, unearthed during excavations under the Vatican in 1950, shows Christ with a solar-rayed halo ascending to heaven in a sun chariot. Some Mithraic ceremonies accompanying the equinoxes and solstices were adopted by Christianity. Christmas can be viewed as a celebration of the winter solstice, particularly since Christ is believed to have been born in the spring.

In India, sun worship was graced by golden-haired gods, among them one with many heads, who rode between heaven and earth on chariots drawn by brilliant horses or horses with many heads. They created with fertilizing warmth and preserved with light, but they also destroyed with burning rays. In Indian mythology, the multiplication of heads and limbs showed increased power. One myth, of interest to later discussions in this book, tells how the sun banished disease from the earth.

In Shinto worship, the Japanese sun goddess, Amaterasu, is the head of a polytheistic pantheon and is also considered to be the ancestor of the imperial family. It is said that Amaterasu sent her son Niniji to pacify the Japanese islands, and that his grandson Jimmu Tenno became the first emperor. Amaterasu is enshrined at Ise, which is also shrine of the nation and of the imperial family. Even today, pilgrims climb Mount Fujiyama and, prostrated, view the sunrise from its summit in homage to Amaterasu, and some Japanese pray to the sun at sunrise and sunset. A rayed sun disk boldly adorns the Japanese flag.

There appears to have been no sun worship in China, although a sketchy and primitive solar mythology is revealed in a few surviving early manuscripts. They described ten suns, each made of fire, that appear in turn in the sky and are carried in a chariot, pulled by dragons and driven by the sun's mother. She bathes the suns in a lake and they ascend a tree. Only the one at the top, the "duty" sun, enters the chariot for the day. When the chariot reaches Mount Yen-tze at the western border, the sun dismounts into a tree and the dragons are unyoked. How the sun gets back east is not explained, nor is a reason given for ten suns. Buddhism and other Oriental philosophical-religious systems emphasizing ethics and introspective development erased much of the earlier sun worship.

Among the American Indians, sun worship took many forms and reached its most highly developed rituals in the most advanced cultures, those of the Mayas, Toltecs, and Aztecs in Central America and the Incas of Peru in South America. These cultures, as in Egypt, identify several sun gods, each representing different aspects of the sun. When the Mayan culture, in which perhaps the earliest development of sun worship occurred, lapsed in the first millennium A.D., sun worship was apparently taken over by other cultures, and passed northward to Mexico and North America and southward to Peru and other parts of South America.

In Central America, an annual sacrifice to the sun was made for the redemption of the world. The young man who volunteered to die was given a year to prepare himself, and after purification and a feast, he was led to the sacrificial alter. Figure 1:7 shows how the priests have removed the young man's quivering heart with a ceremonial knife and are reverently "feeding" it to the sun along with pleas for the welfare of the nation. The murals in the temples in the ancient city of Bonampak in Chiapas, South Mexico, show similar bloodletting and sacrifice of prisoners by Mayan warriors, but it is possible that they depict a later development. Since some cenotes, the limestone pits into which sacrificial victims were hurled, have been found to contain rubber dolls, it may be that at an earlier time symbolic, rather than real, sacrifices were made.

The importance of the sun to these peoples is shown by their concepts of the afterlife. The deceased whose lives had been concerned only with good food and gaiety went temporarily to the Land of Water and Mists and then returned to earth for another cycle. More spiritual beings went to the Land of the Fleshless, and those who had achieved near-perfection went to dwell in the House of the Sun.

FIGURE 1:7. Human sacrifice to the sun as practiced by the Aztecs. From the Codex Florentino as illustrated in Stirling, *Smithsonian Institution Annual Report,* 1946, p. 387, plate 2. Presumably the Mayan sacrifice to the sun was similar to that illustrated here.

The number of human sacrifices to the sun god escalated with the passage of time in Central America. The Aztecs subjugated other tribes and sacrificed prisoners of war to propitiate the sun gods, and apparently some wars were made for the sole purpose of securing victims for sacrifice. The wells into which the bodies of the victims, from whom the hearts had been removed, were hurled are full of human skeletons, and skulls of victims are piled up in the temples.

We know less of sun worship in Peru. The Incas left no written literature, and during the conquest of Peru, the elite were eliminated and the learned priests and their temples destroyed to better enslave the working class. Their traditions and oral literature were soon lost to posterity, but we do know that the Incas tied sun worship to dynastic worship and their rituals were supervised by Incas of high birth and rank. The sun was their tribal god, and the Inca rulers, like the Pharaohs of Egypt, considered themselves "sons" or personifications of the sun god and therefore of divine origin. A highly organized priesthood concerned with sun worship developed elaborate rituals to cover most aspects of life, and rites of confession, penitence, and excommunication were used by the priests to control the population.

The great temples to the sun, built by Incan stonemasons of imbricated many-faced stones with such consummate skill that they have resisted earthquakes, were once decorated with magnificent bejeweled golden disks and other ornaments. Only the few museum pieces that escaped destruction remain to testify to the awe inspired by the sun god and a pantheon of lesser gods.

On the North and South American continents, the slaughter of the elite who controlled sun worship and the subsequent wholesale destruction of entire populations wiped it out, and only certain sun festivals, now in the context of Christianity, are practiced by the remnants of Indian populations in Mexico and in the mountains of South America. The progeny of the conquerors, their slaves, and the new immigrants follow their own kind of worship, chiefly Christianity.

Myths are a people's reflections on existence, their expression of an emotional and intuitive appreciation of the world as they endeavored to explain what physical laws have later elucidated. The myths, poetry, literature, and folk tales that have come down to us from the ancients have outlasted their original purpose but are precious still for their incomparable beauty.

*For Additional Reading**

Edson, L. *Worlds Around the Sun.* New York: American Heritage Publishing, 1969.
Gibson, E. G. *The Quiet Sun.* Washington, D.C.: NASA, 1973.
Halpern, P., Dare, J. V., and Braslau, N. "Sea-Level Solar Radiation in the Biologically Active Spectrum." *Science* 186 (1974): 1204–1208.
Hastings, J., ed. *Encyclopedia of Religion and Ethics,* vol. 12. New York: Charles Scribner's Sons, 1955, pp. 48–103.
*Hawkes, J. *Man and the Sun.* New York: Random House, 1962.
Kiepenheuer, K. *The Sun.* Ann Arbor, Michigan: University of Michigan Press, 1959.
*Menzel, D. H. *Our Sun,* rev. ed. Cambridge, Massachusetts: harvard University Press, 1959.
Myths of the World Series, 18 vols. London: Paul Hamlyn Publishing Group, 1958–1973.
Olcott, W. J. *Sun Lore of All Ages: A Collection of Myths and Legends Concerning the Sun and Its Worship.* New York: Putnam, 1914.
Parker, E. N. "The Sun." *Scientific American* 233 (Sept. 1975): 43–50.
Schramm, D. N. "The Age of the Elements." *Scientific American* 230 (Jan., 1974): 69–77.
Stirling, M. W. "Concepts of the Sun among American Indians." *Smithsonian Institution Annual Report,* 1946, pp. 387-400.
Wainwright, G. "Woodhenges," *Scientific American* 23 (Nov., 1970): 30–38.

*The asterisk preceding an author's name in this and subsequent Additional Reading sections indicates a general reference for those interested in a more extensive report on the topic.

2

Sunlight and Life

Sunlight is a mixture of light visible to us, and ultraviolet and infrared radiation, both invisible to us. Most of the sun's radiation, about 60 percent, is in the infrared region. About 37 percent occurs in the visible region and only about 3 percent in the ultraviolet. In recent years, x-rays and radio waves have been detected from the sun, but the amount of energy represented is vanishingly small. The maximum intensity of sunlight is in the yellow-green part of the spectrum where our eyes see best, presumably an evolutionary adaptation to the sun's radiation.

Visible light, like any white light, is composed of many wavelengths. As the English scientist Sir Isaac Newton demonstrated toward the end of the seventeenth century, light can be dispersed by a prism (or by raindrops) into a rainbowlike spectrum of colors from red (700 nm) at one end to violet (400 nm) at the other. (The nanometer (nm) was recently adopted as the unit of wavelength; a nanometer is one-billionth of a meter, 10^{-9} meter.)

In 1801, Johann Ritter, a German physicist, showed that when sunlight is dispersed by a prism, invisible ultraviolet (UV) rays were present beyond the violet area of the color spectrum. Ritter knew that violet light could reduce silver chloride to silver, and his experiments showed there were invisible rays beyond the violet that could do the same. Invisible infrared rays at the other end of the spectrum, just beyond the red wavelengths, had been discovered by Sir William Herschel, a German-born English astronomer, in 1800 when he found that they could produce heat measured on a thermometer.

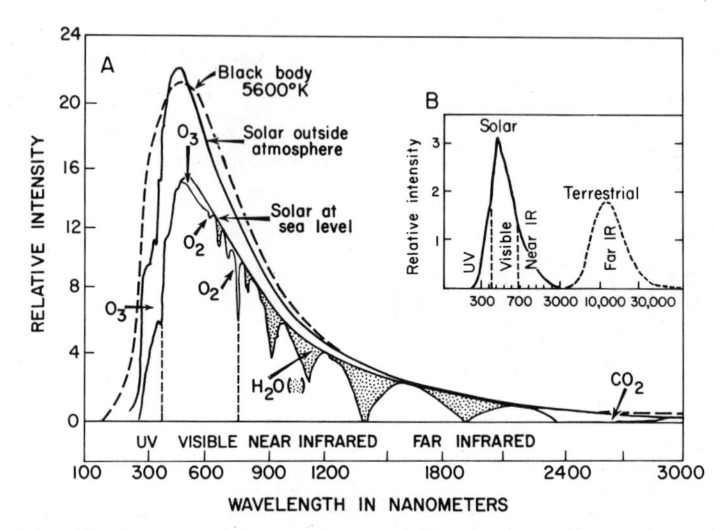

FIGURE 2:1. (A). The sun's spectrum at the outer surface of our atmosphere and at sea level showing absorption of various wavelengths of radiation by atmospheric gases (O_3 = ozone, O_2 = oxygen, H_2O = water, CO_2 = carbon dioxide). The intensity of radiation is given along the vertical scale and the wavelength in nanometers on the horizontal scale. (A nanometer is 1×10^{-9} meter, or one-billionth of a meter). Note absorption of light: by ozone in the ultraviolet (UV) and visible part of the spectrum, by water vapor over most of the infrared (IR) part of the spectrum, and by carbon dioxide in the very long wavelength infrared spectrum. (Intensity of sunlight given in 10 million ergs per square meter per 10 nm bandwidth.) After Hynek, *Astrophysics,* McGraw-Hill, New York, 1951, p. 272. (B). Comparison of extraterrestrial solar radiation with terrestrial radiation in terms of relative intensity of the vertical axis (different scales for sun and earth) plotted against wavelength of radiation on the horizontal axis. Note that the earth's radiation is entirely in the infrared spectrum. The sun is a black body radiator at 5600°K (where K stands for Kelvin or absolute temperature), while the earth is a black body radiator at 250°K. For comparison to something familiar, the ordinary tungsten filament lamp is a radiator at 2760°K. The radiation from the earth is only a small fraction of that from the sun and would not show on the same vertical axis. (An ideal black body emits radiation of all frequencies characteristic of its temperature and absorbs all radiation falling upon it.)

The Ultraviolet Spectrum

At the surface of the earth, ultraviolet radiation from the sun covers a span of the spectrum from 390 to 286 nm. Spectrographic studies of sunlight beyond our atmosphere that were made from rockets and satellites indicate the presence of shorter wavelengths of ultraviolet radiation. As Figure 2:1 shows, when we plot the intensity of each wavelength of the sun's spectrum beyond our atmosphere on a vertical axis against wavelength on a horizontal axis, the radiation from the sun matches a plot of the radiation from a body with a surface temperature of about 5600° absolute (absolute zero = —273.16°C). If a similar spectrographic study of sunlight is made at the earth's surface, it is imme-

Table 2:1. Types of Radiation

Type of radiation	Natural source	Subdivision	Wavelength in nm[a]
Gamma	Radioactive minerals Cosmic rays		0.0001–0.14
X-rays	Sun–low intensity		0.0005–20
		Hard	0.0005–0.1
		Soft	0.1–20
Ultraviolet[b]	Sun		40–390
		UV-C	40–286
		UV-B	286–320
		UV-A	320–390
Visible	Sun		390–780
		Violet	390–430
		Blue	430–470
		Blue-green	470–500
		Green	500–530
		Yellow-green	530–560
		Yellow	560–590
		Orange	590–620
		Red	620–780
Infrared	Sun		$780–4 \times 10^5$
		Near	$780–2 \times 10^3$
		Far	$2 \times 10^3–4 \times 10^5$
Hertzian waves	Sun–low intensity		$10^5–3 \times 10^{13}$
		Space heating	$10^5–10^6$
		Radio	$10^6–10^{12}$
		Radar	$10^6–10^9$
		Television	$10^9–10^{11}$
Power A.C.	?		10^{15}

[a] 1 nm (nanometer) = 10^{-9} meter. 1 nm = 1 mμ = 10 Å (angstrom). Although various units of length are used to measure wavelength of light, the nanometer is recommended.
[b] These subdivisions of the ultraviolet spectrum are arbitrary, but they facilitate a simpler presentation of the subject matter of this book. Even these subdivisions are not always given the same limits; thus, the originator of these subdivisions, Coblentz (*Journal of the American Medical Association* 99 [1943]: 125) gave UV-A as 400–315 nm; UV-B as 315–280 nm, and UV-C as less than 280 nm. Photobiologists prefer the subdivisions: near-UV radiation, 390–310 nm; far-UV radiation, 310–200 nm, and vacuum-UV radiation, 200–40 nm; their use would be cumbersome here.

diately apparent that the shorter wavelengths of ultraviolet radiation are missing. Evidently, short-wavelength ultraviolet radiation is absorbed by parts of the earth's atmosphere. Various wavelengths in the visible and infrared portions of the spectrum are also absorbed to various degrees.

In discussing the biological effects of the ultraviolet portion of the sun's spectrum, we divide it into three regions: ultraviolet A (UV-A), ultraviolet B (UV-B), and ultraviolet C (UV-C) radiation. Table 2:1 presents the various kinds of radiation from the sun and indicates the three types of ultraviolet rays. Ultraviolet A radiation extends from 390 to 320 nm, just beyond the violet portion of the visible spectrum. UV-A has sometimes been called biotic UV radiation and was once considered to be beneficial. Now we know that in large doses it may be quite damaging. Ultraviolet B extends from 320 to 286 nm, the latter wavelength being the short wavelength limit of sunlight at the surface of the earth. Among other things, UV-B damages the skin, causing sunburn. Ultraviolet C includes wavelengths shorter than 286 nm and overlaps the x-ray region at 40 nm. UV-C radiation in sunlight impinges on the top surface of the earth's atmosphere and is absorbed by ozone. It does not reach the earth's surface. Highly destructive of life (biocidal), UV-C radiation may be produced in the laboratory and has been used to induce hereditary changes in organisms (mutations), to study repair of nucleic acids damaged by it, to treat the skin, and to sterilize surfaces.

Ordinary window glass transmits UV-A radiation but absorbs UV-B and UV-C rays. Quartz, on the other hand, transmits all three types, including UV-C radiation, to about 200 nm. That is why quartz lenses, prisms, and dishes are used to study the effects of UV-B and UV-C radiation. Other lenses, prisms, and culture dishes must be used for studies with wavelengths shorter than 200 nm. Since oxygen absorbs such radiations and produces ozone highly toxic to life, these experiments must be performed with spores or cysts and other dry forms of life in an inert gas such as nitrogen, or in a vacuum.

How the Atmosphere Absorbs UV-C Radiation. Unpolluted atmospheric air at the surface of the earth consists of 78.1 percent nitrogen (N_2), 20.9 percent oxygen (O_2), 0.033 percent carbon dioxide (CO_2), 0.9 percent argon (A), and very small amounts of neon (Ne), helium (He), krypton (Kr), xenon (Xe), hydrogen (H_2), methane (CH_4), and nitrous oxide (N_2O). Samples of air taken with balloons and rockets at progressively higher levels of the atmosphere indicate that the relative proportions of the various gases in air remain the same, while the con-

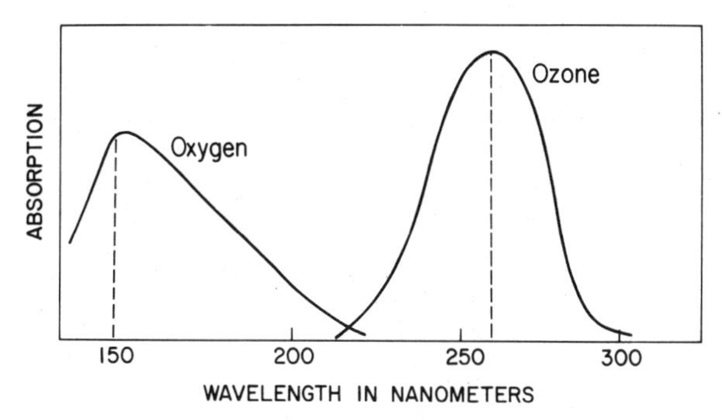

FIGURE 2:2. Diagrammatic representation of the ultraviolet absorption bands of molecular oxygen and ozone. After Duvalier, *The Photochemical Origin of Life,* Academic Press, New York, 1965, p. 96.

centrations decrease with height. However, when samples of air are taken between 15 and 35 kilometers above the earth's surface, ozone (O_3), another gas, is found. In midlatitudes, ozone reaches its highest concentration at about 25 kilometers. This layer of the atmosphere, between 15 and 35 kilometers, is therefore called the ozone layer.

When the spectrum of sunlight is measured at various atmospheric levels, we discover that light passing through the ozone layer removes all the UV-C radiation—from the shortest to wavelength 286 nm, the shortest ultraviolet wavelength detected at the earth's surface. Occasionally, unusual changes in the ozone layer permit momentary transmission of shorter ultraviolet wavelengths.

From laboratory experiments we learn that oxygen absorbs the very short wavelength UV-C radiation with peak absorption at 150 nm, its absorption decreasing with increasing wavelength. As Figure 2:2 shows, ozone, on the other hand, absorbs the remaining UV-C radiation, with peak absorption at 260 nm. When oxygen absorbs UV-C-type rays high in the atmosphere, it forms ozone. And, as the following reaction describes, when ozone itself absorbs somewhat longer UV-C rays, some of it decomposes:

$$O_2 \xrightarrow[\text{peak 150 nm}]{\text{UV-C radiation}} 2O$$

$$O + O_2 \longrightarrow O_3$$

$$O_3 \xrightarrow[\text{peak 260 nm}]{\text{UV-C radiation}} O_2 + O$$

Remember that O_2 is molecular oxygen, O_3 is ozone, and O is atomic oxygen. Much of the atomic oxygen so produced recombines with molecular oxygen to form ozone as in the second reaction, above. The energy of the light quanta in these reactions is released as heat. As expected, the ozone layer is warmer than the layers immediately above and below it.

Because UV-C radiation striking the upper surface of our atmosphere is biocidal, the ozone "umbrella" makes life possible in water and on the surface of the land. Without this protection, living things on earth would be damaged or killed.

The total amount of ozone in the atmosphere is relatively small. If the ozone layer were placed under what we call standard conditions of one atmosphere pressure—15 pounds per square inch, equivalent to a "thickness" of 760 mm mercury—and at 0°C, it would have a thickness (or partial pressure) of 2.4 to 4.6 mm, or the equivalent of about 0.00605 atmospheric pressure. Under the same conditions, oxygen in the atmosphere has a thickness of about 160 mm or 0.2099 of atmospheric pressure. The ozone in the atmosphere varies seasonally from a thickness of 2.4 to 2.6 mm at the equator and 3.1 to 4.3 mm at 70°N latitude. Ozone also varies with sunspot activity, during which the sun rains charged particles and electric storms affect our atmosphere. Smaller ozone changes also occur locally during the course of each day.

Clearly, our protective ozone layer depends on the presence of oxygen from which it is formed, and the ozone layer exists only because our atmosphere contains oxygen. If earth's primitive atmosphere did not have appreciable amounts of oxygen, it would not have had an ozone layer. Without that cover, considerable UV-C radiation—in addition to the other types of UV rays—would have reached the earth's surface. Without ozone, around 4.6 billion years ago, life as we know it would not have been possible on the surface of the earth.

UV-B Radiation. As ozone concentration changes, the intensity of ultraviolet radiation reaching the earth also changes. Rarely do we receive shorter wavelengths below the 286 nm limit, but the intensity of the UV-B radiation increases as ozone levels diminish and B-type rays decrease as ozone builds up.

On the earth's surface today, UV-B rays are of very low intensity compared to UV-A radiation. Keep in mind, however, that UV-B radiation is more damaging to life than UV-A. Doses of UV-A radiation two or more orders of magnitude (100 to 1000X) greater are required to produce damage equivalent to that produced by UV-B radiation, except when a photosensitizer is present (see Chapter 7).

Scattering, in addition to absorption, attenuates the intensity of sunlight. Scatter is inversely proportional to the fourth power of a wavelength and it is therefore greatest for the shortest wavelengths. Obviously, UV-B radiation is subject to the greatest atmospheric scattering. Clouds and fog scatter ultraviolet radiation and absorb little of it, but they do absorb infrared radiation, thus cooling the effect of sunlight.

The Primitive Atmosphere

We take our atmosphere for granted, but other celestial bodies are not as fortunate. Atmospheres exist only if there is a gravitational field capable of holding gases. For example, the moon's gravity is not strong enough to maintain an atmosphere, even though during volcanic activity gases must have come from the moon's interior. The sun, with its enormous gravitational field, can contain even the lightest gas, hydrogen. Not even the earth can hold onto that gas, but as we know, the earth's gravitational field is sufficient to hold onto the gases that characterize our present atmosphere.

The earth's atmosphere was not always like it is now. Some scientists believe that all of our solar system was formed at one time from the gaseous material of a solar nebula. Measurements of radioactive decay in rocks indicate that this event probably occurred about 4.6 billion years ago. The primary atmosphere of the newborn earth probably resembled that of the sun from which it arose and, like the sun's atmosphere, contained hydrogen, helium, neon, argon, krypton, and xenon in large amounts. When we compare the amounts of these elements present today on earth and on the sun—assuming that our sun has about the same atmosphere that it had at the time of the birth of the planets—we discover striking discrepancies. As Table 2:2 reveals, using the nonvolatiles silicon, magnesium, and aluminum as a basis for a comparison of our solar system and the earth, these gaseous elements differ by several orders of magnitude.

Most, if not all, the light gases in the original atmosphere must have left the earth, probably because of its weak gravitational field, and

Table 2:2. Relative Amounts of Elements on Earth and in the Solar System in Atoms per 10,000 Atoms of Silicon[a]

Element	Whole earth	Solar system	Approximate ratio, solar system to earth
Hydrogen	250	260,000,000	1 million
Helium	0.00000035	21,000,000	60 million million
Carbon	14	135,000	10 thousand
Nitrogen	0.21	24,400	1 hundred thousand
Oxygen	35,000	236,000	10
Neon	0.0000012	23,000	20 thousand million
Sodium	460	632	1.4
Magnesium	8,900	10,500	1.2
Aluminum	940	851	0.9
Silicon	10,000	10,000	Basis of comparison
Argon	0.00059	2,280	4 million
Krypton	0.00000006	0.69	11.5 million
Xenon	0.000000005	0.07	14 million

[a] After Rasool, *Science* 57 (1967): 1466.

its high temperature accelerated diffusion of the gases. Those light gases that still remain from the original atmosphere continue to leave the earth by diffusion. Concentrations of gases in our atmosphere result from an equilibrium between their gain—from the interior of the earth and from the splitting of water into hydrogen and oxygen by UV-C radiation high in the atmosphere—and from their loss by diffusion.

Life and the Secondary Atmosphere

We can presume that a secondary atmosphere appeared only when the surface of the earth had cooled down. Thought to have been anaerobic—that is, lacking in oxygen—the secondary atmosphere accumulated slowly about the cooled earth from gases emanating from the interior of the earth, out of volcanoes and fumaroles (small holes from which volcanic vapors escape). Today, gases from volcanoes and fumaroles continue to escape, but at a much lower rate than during the times of cataclysmic change that occurred periodically in the earth's long history.

The exact composition of the secondary atmosphere in the anaerobic period is the subject of much study. Some geologists think that analyses of the present outpourings from volcanoes and fumaroles, as pre-

Table 2:3. Composition of Gases Issuing from Volcanoes, Fumaroles, Steam Wells, Geysers, and Rocks. Median Analyses, Recalculated from Volume to Weight Percentages[a]

Substance	Chemical symbol	Kilauea and Mauna Loa volcanoes	Fumaroles, steam wells, and geysers	Basalt and diabase rocks[b]	Obsidian, andesite, and granite rocks[b]
Water	H_2O	57.8	99.4	69.1	85.6
Carbon dioxide	Total C as CO_2	23.5	0.33	16.8	5.7
Sulfur	S_2	12.6	0.03	3.3	0.7
Nitrogen	N_2	5.7	0.05	2.6	1.7
Argon	A	0.3	tr	tr	tr
Chlorine	Cl_2	0.1	0.12	1.5	1.9
Fluorine	F_2	–	0.03	6.6	4.4
Hydrogen	H_2	0.04	0.05	0.1	0.04

[a] After Rubey, *Bulletin of the American Geological Society* 62 (1951): 1137.
[b] Rocks produced from molten magma either inside the earth (crystalline) or outside, some glassy like obsidian, others crystalline like granite.

Table 2:4. Suggested Envelopes for Primitive Earth[a]

Sphere	Constituents	Author
Atmosphere	CH_4, NH_3, H_2O, H_2[b]	Oparin, Urey
Atmosphere	$CH_4 \rightarrow CO_2$, $NH_3 \rightarrow N_2$, H_2O, H_2	Bernal
Hydrosphere	CO_2, NH_3, H_2S, H_2O	Bernal
Atmosphere	CO_2, N_2, H_2S, H_2O	Rubey
Hydrosphere	CO_2, NH_3, H_2S, H_2O	Rubey
Atmosphere	CO, CO_2, N_2, H_2S, H_2O	Revelle

[a] Key to symbols: CH_4 methane; NH_3 ammonia; H_2O water; H_2 hydrogen; CO_2 carbon dioxide; N_2 nitrogen; H_2S hydrogen sulfide; CO carbon monoxide.
[b] From Fox, "The Chemical Problems of Spontaneous Generation" *Journal of Chemical Education* 34 (1957): 473.

sented in Table 2:3, offer the best indicators of its probable nature. But not everyone agrees. Others postulate quite different constituents (see Table 2:4). For example, one group suggests that the atmosphere in this next period was composed of methane, ammonia, hydrogen, and water vapor, while another group suggests carbon dioxide, carbon monoxide, hydrogen, nitrogen, and water vapor.

From all this, however, what is of most interest is whether water vapor and oxygen existed. It is agreed that in the earth's secondary atmosphere water vapor was present and that free oxygen, if present, occurred in only extremely small amounts. In other words, when the secondary atmosphere appeared as the earth cooled down, it was essentially a reducing atmosphere—one in which the oxygen had combined with some of the gases to form new compounds. One of the many reasons for this conclusion is that most ancient rocks and deposits of minerals immediately combine with oxygen upon exposure to air.

When water vapor is present in an atmosphere, a small amount will be split into hydrogen and oxygen because it absorbs UV-C radiation, with a peak at about 175 nm. How UV-C radiation passed through the secondary atmosphere depended on what gases were present, and it is likely that water vapor at almost all levels of the atmosphere could have been split into hydrogen and oxygen. Some hydrogen would have diffused into space. The residual oxygen could not have accumulated as the water decomposed, because once it had reached about 0.0001 of the level of oxygen present in the atmosphere today, it would absorb the very radiation that splits water (Figure 2:2). As small amounts of oxygen combined with other elements or compounds, more would appear from the splitting of water by light, and the equilibrium concentration would remain at 0.0001 of the present atmospheric level (PAL).

While small amounts of ozone would form from oxygen even at 0.0001 of the present atmospheric level, its absorption of UV-C radiation would be too slight to screen the earth's surface. The secondary atmosphere, whatever its composition, probably permitted considerable UV-C rays to reach the earth's surface.

Those interested in the origin of life believe that UV-C radiation was a major source of energy in biochemical evolution. Before life could begin on earth, many inorganic and organic compounds had to be synthesized, including high concentrations of organic compounds found in living cells. It took perhaps a billion years for these to develop. For life to appear, these compounds had to combine into a primitive self-replicating unit capable of using other organic materials in the environment in order to duplicate and grow. Other possible sources of energy for biochemical evolution are ionizing radiations, electric storms, and heat.

Geological evidence suggests that the earth's surface was without life for a long time, probably because of the devastating effects of UV-C radiation reaching the surface of the primitive earth. To survive, life in its early stages must have been nurtured in murky waters or in the shade of opaque objects to escape bombardment from UV-C rays. But UV-C radiation must also have served in the development of early life, because it greatly increases the rate of mutation. Those life units that survived exposure to the radiation must have acquired valuable mutations in the struggle for survival.

All early life was anaerobic, because the atmosphere lacked oxygen for aerobic existence. It is likely that the waters of the primitive earth, before life began, were organic soups synthesized from available inorganic matter mainly by the action of UV-C radiation and heat. As nutrients in fresh waters and oceans declined, mutants must have appeared that were capable of synthesizing some of the organic components they had previously obtained from solution in the water. Some believe that the ability to synthesize certain organic molecules in anaerobic life must have happened as a result of natural selection at the same time that nutrients in the earth's water became exhausted. According to the heterotroph hypothesis, living units must have then appeared that could survive on a single organic source of food, synthesizing all the other things they needed. Living units that thrive on organic matter obtained from their surrounding medium are called heterotrophs.

Evolution of Photosynthetic Cells and Development of an Ozone Layer

Heterotrophic organisms require organic nutrients to grow and cannot use the sun's energy to maintain their life. Obviously, organisms that could use energy from sunlight to synthesize their own organic materials from inorganic sources abundant in their surroundings would have had great survival power in an environment depleted of organic nutrients. The development of phototrophs that could use the sun's energy to produce organic from inorganic nutrients was therefore an essential step in the evolution of life.

In purple sulfur bacteria:

$$CO_2 + 2H_2S \longrightarrow 2S + H_2O + (CH_2O)$$

$$2CO_2 + H_2S + 2H_2O \longrightarrow H_2SO_4 + 2(CH_2O)$$

In green plants:

$$CO_2 + 2H_2{}^{18}O \longrightarrow {}^{18}O_2 + (CH_2O)$$

FIGURE 2:3. Reactions in photosynthesis of bacteria and green plants. In the early stages of a culture of the photosynthetic sulfur bacteria, hydrogen sulfide (H_2S) is oxidized and sulfur (S) is deposited in the bacteria. But, when the supply of H_2S has been used up, the sulfur in the bacteria is oxidized to sulfuric acid (H_2SO_4). CH_2O represents cell material, not necessarily sugar or other carbohydrate. In photosynthetic plants (algae and higher plants), it can be demonstrated that water (H_2O) is the source of hydrogen for reducing carbon dioxide (CO_2) by using water labeled with a heavier isotope of oxygen (^{18}O) instead of the common oxygen (^{16}O). When this is done, the labeled oxygen (^{18}O) comes out only in the gaseous oxygen ($^{18}O_2$). The main difference between photosynthesis in plants and in bacteria is the source of hydrogen for reducing carbon dioxide. Water (H_2O) is used in the oxidation of hydrogen sulfide to sulfuric acid by the photosynthetic bacteria but oxygen is not formed as a byproduct.

Fossil photosynthetic cells two billion years old have recently been discovered. Photosynthesis is the process by which inorganic carbon turns into organic carbon by using the sun's energy. It is not known whether the several types of aquatic photosynthetic organisms developed at the same time or in sequence. It is possible, however, that anaerobic aquatic photosynthetic cells appeared first in the sequence.

Anaerobic photosynthetic bacteria exist today. They contain chlorophyll that fixes carbon dioxide using light energy of the sun. The purple sulfur bacteria reaction shown in Figure 2:3 indicates how they get the hydrogen needed for the reduction of carbon dioxide from such compounds as hydrogen sulfide and produce sulfur (or sulfur compounds) as oxidation products. Their "bacteriochlorophyll" is photosynthetically active and differs in chemical structure from plant chlorophyll only in detail, and they absorb visible light at wavelengths longer and shorter than those absorbed by plant chlorophyll. Since, in order to reduce carbon dioxide, they use hydrogen sulfide (or organic) donors of hydrogen, which are not abundant in nature except as they appear from organisms ultimately dependent upon photosynthesis, they occupy a restricted place in the economy of nature.

At one point in the history of the earth, mutants of early living units appeared that could synthesize carbohydrates from carbon dioxide and water through photosynthesis, with water serving as the source of hydrogen for reducing carbon dioxide—the most profound step in the evolution of life as we know it. Not only did these mutants synthesize their organic materials but they also liberated oxygen into the atmosphere, as in the green plant reaction given in Figure 2:3. In this striking new development, these organisms performed several critical operations for the future of life. Not only did they manufacture their own organic nutrients required for the heterotrophic organisms, but they also produced oxygen for the more efficient use of nutrients in the metabolism of aerobic cells. What is more, the oxygen they provided also formed the ozone screen that ultimately protected life from UV-C radiation.

What were these original oxygen-producing photosynthetic life units like? They may have been similar to today's blue-green algae, which are much like bacteria in the simplicity of their cellular organization, but also carry out much the same photosynthetic processes as the more highly organized green algae and higher plants. Blue-green algae can fix inorganic carbon dioxide of the air into organic compounds using the energy of absorbed sunlight. In addition, they fix nitrogen; that is, they can live in a medium without nitrogen "fixed" in nitrates or ammonia in their surroundings, because they can reduce nitrogen from the air to an ammonialike compound for use in synthesis of their own proteins. It is not known whether blue-green algae, without a distinct cell nucleus, gave rise to green algae, which have a distinct nucleus within a nuclear envelope—a radical step in nuclear organization characteristic of the cells of higher organisms.

Perhaps later these primitive oxygen-producing photosynthetic life units became differentiated into a number of patterns foreshadowing the green, red, and brown algae and green plants of today. Like the algae in the fossil records, some present-day algae are microscopic, and others—the giant kelps, for example—are very large indeed. But the basic photosynthetic mechanism is very much the same in all such forms of life.

Life on land first appears in fossil records early in the Phanerozoic eon, some 400 million years ago (Figure 2:4). Subsequently, terrestrial plants—liverworts, mosses, ferns, coniferous plants, and finally, flowering plants—evolved in succession. They covered the land rapidly and during favorable climatic periods plant life on earth was lush indeed. The coal- and oil-bearing layers below the earth's surface offer us their remains. It is likely that during that period atmospheric oxygen reached our present level, or possibly even surpassed it severalfold. Oxygen in the

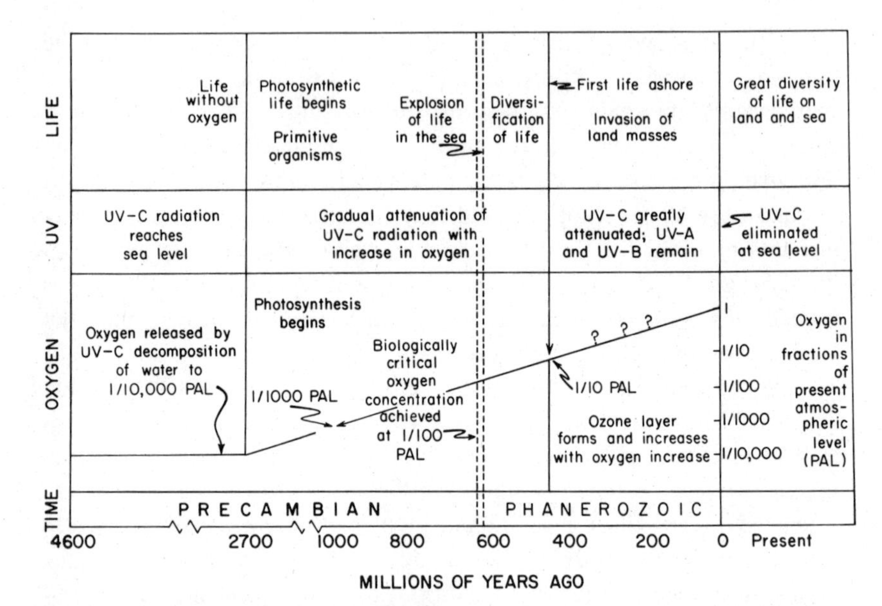

FIGURE 2:4. The relation between availability of oxygen and the evolution of life in the sea and on land, as suggested by fossils and correlations with deposits in minerals. The approximate timing is based on radioactive decomposition products in the rocks and fossils. Not all workers in the field agree with the details presented here from data in Berkner and Marshall, *Journal of Atmospheric Science* 22 (1965), p. 257. The oldest fossil cells are now dated at 3.4 billion years, the oldest photosynthetic cells at 2 billion years ago.

atmosphere probably reached its present level by the middle of the Phanerozoic eon.

Life in the Aerobic Phase

Only long after photosynthetic organisms appeared was it possible to achieve our current level of atmospheric oxygen. At first, extreme localization of such living forms prevented large-scale accumulation of oxygen. As they slowly spread over the surface of the earth's waters, these plants produced ever-increasing amounts of oxygen. Fossil records are our key to the major geological events in the earth's history, and enable us to postulate stages in the achievement of the present atmospheric oxygen levels.

The sudden appearance of all the invertebrate types (phyla) in the fossil record, for example, at the very beginning of the Phanerozoic Eon, or about 600 million years ago, suggests that some very important changes occurred in geological conditions. While there is no strict agreement with Lloyd Berkner and Lauriston Marshall, two American geologists, they argue that this was the period when the first biologically critical level of oxygen (0.01 of the present atmospheric level) was achieved.* This level is the partial pressure of oxygen at which respiration (aerobic oxidation) supersedes fermentation (anaerobic oxidation). Its development would have permitted aerobic respiration in cells and offered a more efficient use of energy stored in nutrients to maintain life and growth. According to Nobel laureate George Wald, "Fermentation is so profligate a way of life that photosynthesis could do little more than keep up with it. To use an economic analogy, photosynthesis brought organisms to the subsistence level; respiration provided them with capital. It is mainly this capital that they invested in the great enterprise of organic evolution."

Fossil records show that all early life—plant, animal, or microbial—appears to have been aquatic and presumably life originated in the sea. Before the protective ozone layer covered the earth, life could not have resisted the lethal ultraviolet radiation on land.

There is very little information about how UV-C and UV-B rays transmit through sea water because oceanographic measurements of sunlight into sea water usually terminate at wavelength 350 or 330 nm. Available data, however, suggest that thin layers of clear sea water transmit UV-C rays up to a wavelength of about 280 nm, and that a meter of clear sea water will absorb even the UV-B radiation almost completely at 290 nm. When the water contains organic material, such as is assumed to have been present in the "soups" of the primitive earth, transmission of short wavelength ultraviolet radiation would probably have been slight.

Early life before the ozone layer probably lived in the sea at depths where lethal radiation could not have penetrated or behind opaque objects in shallow water. However, when photosynthetic organisms appeared, they had to live in water at shallower depths to enable them to get sufficient light for photosynthesis.

While much of the study of the origin of life and the nature of atmospheric development is still speculative, most scientists agree about one thing. Ozone in our atmosphere is essential to the protection and continuation of life on earth as we know it.

*Fossil evidence of true cells likely to have required oxygen is found in rocks 1.2 billion years old by tracer chronology.

For Additional Reading

*Berkner, L. V. and Marshall, L. C. "The History of Growth of Oxygen in the Earth's Atmosphere." In *The Origin and Evolution of Atmospheres and Oceans,* edited by P.F. Brancazio and A. G. W. Cameron. New York: John Wiley and Sons, Inc., 1964, pp. 86–101.

Brinkmann, R. T. "Dissociation of Water Vapor and the Evolution of Oxygen in the Terrestrial Atmosphere." *Journal of Geophysical Research* 74 (1969): 5355–5358.

Calvin, M. "Evolution of Photosynthetic Mechanisms." *Perspectives in Biology and Medicine* 5 (1962): 147–172.

Cloud, P. "Evolution of Ecosystems." *American Scientist* 62 (1974): 54–66.

Cloud, P. and Gibor, A. "The Oxygen Cycle." *Scientific American* 223 (Sept., 1970): 110-123.

Fischer, A. G. "Fossils, Early Life and Atmospheric History." Proceedings of the National Academy of Science, U.S. 53 (1965): 1205–1215.

Gates, D. M. "Spectral Distribution of Solar Radiation at the Earth's Surface." *Science* 151 (1966): 523–529.

Miller, S. L. *The Origins of Life on the Earth.* Englewood Cliffs, N.J.: Prentice Hall, 1973.

Morgulis, L. *Origin of Eukaryotic Cells.* New Haven, Conn.: Yale University Press, 1970, 349p.

Ponnamperuma, C. "Ultraviolet Radiation and the Origin of Life." In *Photophysiology,* vol. 3, edited by A. C. Giese. New York: Academic Press, 1968, pp. 253–267.

*——. *Exobiology.* Amsterdam: North Holland Publishing Co., 1972, pp. 1–15, 16–61, 95–135, 369–399.

——. *Origins of Life.* New York: Dutton, 1972, 215p.

Rubey, W. W. "Geologic History of Sea Water: An Attempt to State the Problem." In *The Origin and Evolution of Atmospheres and Oceans,* edited by P. J. Brancazio and A. G. W. Cameron. New York: John Wiley and Sons, Inc., 1964, pp. 1–63.

Schopf, J. W., Haugh, B. N., Molnar, R. E., and Sattertwait, D. F. "On the Development of Metaphytes and Metazoans." *Journal of Palaontology* 47 (1973): 1–9.

Siever, R. "The Earth." *Scientific American* 223 (Sept., 1975): 82–90.

Smith, R. C. and J. E. Tyler "Transmission of Solar Radiation into Natural Waters." In *Photochemical and Photobiological Reviews,* vol. 1, edited by K. C. Smith. New York: Plenum Publishing Corp., 1976.

Stanley, S. M. "An Ecological Theory for the Sudden Origin of Multicellular Life in the Late Precambrian." Proceedings of the National Academy of Science, U.S. 70 (1973): 1486–1489.

Urbach, F., ed. *The Biologic Effects of Ultraviolet Radiations.* Oxford: Pergamon Press, 1969, pp. 329–333, 335–339, 377–390.

Wald, G. "The Origin of Life." *Scientific American* 191 (Aug. 1954): 45–53.

3

Is Sunlight Good for You?

The conventional wisdom that the sun cures human illness comes to us from ancient times and persists to this day. Peasants in some parts of southern Europe expose sick people and sick animals to the sun in hopes of a cure. Most people believe that sunlight is highly beneficial to health, and they expose themselves to it whenever they can—on holidays, at beaches, or under sun lamps.

Sunbathing

Sunbathing is not a new custom; it goes back at least two thousand years to Greek and Roman times when enclosed patio solaria open to the sun were quite popular. The Greek historian Herodotus (484?–425? B.C.) counseled, "Exposure to the sun is eminently necessary to those who are in need of building themselves and putting on weight." Physicians in medieval Arabia also proposed sunbaths for general well-being.

Even recent reports suggest that some good may come from sunlight on nerve endings, echoing what Herodotus recommended to the ancient Greeks: "It is principally the back that should be exposed to the sun . . . as, if the nerves obeying the will are maintained in a state of mild heat, the whole body will benefit by it. However, the head must be protected by a blanket." Mayer, an American physician, concludes that whatever one may think about all this, if the effect of sunlight "is more general than specific, it is not one bit less important for that reason."

Reports on the general health benefits of sunlight persist and some try to pinpoint specific ways in which the sun is helpful. Since window glass blocks middle-wavelength ultraviolet from passing through and thus prevents sunlight from affecting us, it has therefore been assumed that it is the sun's ultraviolet rays that are most effective, specifically the UV-B radiation.

According to the German scientist Seidl, experiments in northern countries such as Sweden and Russia show statistically significant benefits from exposure to artificial sources of ultraviolet radiation in classrooms and factories. Reports indicate improved performance on an exercise-tolerance test, increased immunological responsiveness as measured by challenges with antigens, and reduced respiratory infections. It has been suggested that ultraviolet radiation does remarkable things: It lowers blood pressure, lowers blood cholesterol, increases blood hemoglobin, improves cardiovascular work capacity (perhaps resulting in better circulation), stimulates nerve endings in the skin, promotes vitamin D synthesis and, in general, increases physical fitness. Breathing is supposedly improved, especially in asthmatics. Unfortunately, not all experimenters have been able to repeat these results and one hopes for more controlled studies to learn whether these claims have merit. Perhaps there are benefits to be expected for people in northern climates who find little sunshine in their long winters. Those from southern climates may already receive such good fortune from the sun.

Bacteria and Ultraviolet Radiation

In their pioneering experiments in 1877, Downes and Blunt showed that sunlight kills bacteria. Soon after, it was learned that the ultraviolet rays in sunlight were responsible, and following this principle, today's hospitals use ultraviolet light to sterilize air and in operating rooms. In the laboratory it is routinely used to sterilize research culture chambers.

Because sunlight inactivates viruses and kills bacteria, including pathogenic (disease-producing) varieties, it probably helps reduce the spread of disease when our skin, clothes, and utensils are exposed to it. Ultraviolet radiation works on the ground and in the water as well as on the surface of organisms. Because bacteria and viruses are small, sunlight easily reaches their vital genetic material. It takes much longer for the sun to affect animal and plant cells, because more cytoplasm is interposed between the surface of the cell and the condensed genetic material in the centrally located nucleus. Viruses and bacteria on the surfaces of plants and animals can probably be inactivated without much damage to the underlying cells. In this way, sunlight ultraviolet radiation can be quite beneficial to man, animals, and plants.

But ultraviolet radiation has serious limitations in its use against viruses, bacteria, and other pathogens because of its shallow penetration in tissues, nutrient-rich natural waters, and opaque objects. If the pathogens are layered or covered with slime or dirt, only the most superficial ones will be inactivated. Ultraviolet radiation cannot replace chemical bactericidal agents and is best used together with chemicals or to maintain sterile surfaces already treated chemically.

Skin Tuberculosis

Lupus vulgaris, tuberculosis of the skin, once occurred widely in northern Europe, where sunlight is weak or absent for long periods in late autumn, winter, and early spring. Ulcers would develop on the face and neck that, even when cured, left ugly scars.

Niels Finsen (1860–1904), a Danish physician, demonstrated that sunlight concentrated by quartz lenses (which transmit UV-B radiation) on the ulcers effected dramatic cures. Finsen worked with an engineer, himself afflicted with lupus, who designed a carbon arc that would serve as a steady source of ultraviolet radiation in the range of wavelengths desired for the cure. Using a quartz flask filled with water (to remove the heat but transmit the UV-B radiation) placed between the arc and the patient, and a quartz lens to focus the light, Finsen achieved remarkable cures. The grateful people of Copenhagen built him an institute bearing his name to study the effects of light on organisms (Figure 3:1). Apart from his scientific research, Finsen also continued clinical studies, especially on lupus. He found that illuminating the whole body with sunlight accelerated the cure. While he could not prove it, he thought that ultraviolet radiation killed bacteria in lupus lesions. Since UV-B radiation is readily absorbed by cells, it could not possibly reach the bacteria in all the affected cells. It is now thought that UV-B induces an inflammation that stimulates immunological reactions in the tissue, subsequently healing the lesion. In 1903 Finsen received the Nobel Prize in medicine for his achievement.

FIGURE 3:1. Finsen light therapy, from a painting by Krøyer. Finsen is the individual in the white gown in the middle of the group of three discussants on the left side of the picture. By permission of the Danish Foreign Office (Journal Publication #34, September, 1960), and courtesy of Dr. Holger Brodthagen, Finsen Medical Institute, Copenhagen.

Lupus has virtually disappeared, but occasionally occurs in Eastern Europe's northern areas. Lupus is now treated with antibiotics, such as isoniazid and streptomycin, to which it is quite responsive. However, ultraviolet radiation therapy is still administered together with antibiotic chemotherapy for those cases unresponsive to chemotherapy alone.

Bone and Pulmonary Tuberculosis

At one time—actually not long ago—much was made of the curative effects of exposure to sunlight of both bone and pulmonary tuberculosis. Patients flocked to sanatoria in the mountains or by the sea, far from big cities, where they spent many hours in the sun. It was believed that this therapy was effective because of exposure to the ultraviolet radiation from the sun. However, at present, it is thought that these sanatoria were effective because patients breathed air unpolluted by irritants in industrial and automobile smog, and enjoyed the benefits of relaxed and pleasant surroundings. To this day we still are not sure whether the sun's ultraviolet radiation plays a role in curing tuberculosis, and strangely, despite the lack of convincing proof, not everyone has abandoned the view that the sun's rays have such curative value. At present, chemotherapy—rifamycin, isoniazid, streptomycin, etc.—cures tuberculosis so effectively that physicians have lost interest in studies on ultraviolet therapy.

Skin Diseases

The list of skin diseases once treated with ultraviolet radiation fills a whole page in a recent historical account. Various skin diseases were treated by using ultraviolet radiation from carbon and mercury arcs with quartz optics. Yet, how these cures were achieved has never been fully understood. Even in those cases where bacteria caused the disease, no one has demonstrated that the bacteria themselves were killed by exposure. More likely, as in lupus vulgaris, in these diseases inflammation or its consequences stimulate the skin to enact a cure. Today, antibiotics and other chemical and drug therapies supplant the cumbersome use of ultraviolet radiation therapy.

In some cases, however, ultraviolet radiation continues to be used in conjunction with antibiotics and drugs to treat a few skin diseases. Until recently UV-B rays were still used to treat psoriasis, in which silvery scales are formed in rings on the skin. Still, no one yet knows how these rays support the action of the drugs. Occasionally, acne, infection of hair follicles, and boils are also treated with ultraviolet radiation, as are certain sensitivities to light. In Chapter 7 we will consider the successful repigmentation of white spots on skin of all colors and long-term remission from the itching of psoriasis by photosensitization of the skin to UV-A radiation by psoralens extracted from plants, the treatment of herpes virus in humans by photodynamic dyes and visible and UV-A rays, and of superficial cancers in mice by photodynamic dyes and laser beams.

Vitamin D and Rickets

Rickets, the disease found especially among infants and children, is caused by a disturbance in normal bone formation and hardening. The long bones bend, causing bowlegs, and muscular contraction is distorted. Nodular enlargements form at the ends and sides of bones, the "soft spot'' on the heads of infants closes slowly, muscles ache, and ultimately the liver and spleen degenerate. Rickets, the first air-pollution disease, was prevalent among children in northern European industrial towns, where the burning of soft coal produced heavy black smog along the narrow streets and UV-B antirachitic radiation from sunlight was greatly reduced. We see many deformed subjects in paintings from the seventeenth and eighteenth centuries who were probably victims of rickets. Indian and Pakistani children in British industrial cities suffered even more than British youngsters, because the dark pigment known as melanin, present in greater abundance in their skins, cut off much of the remaining antirachitic UV-B radiation.

Humans are not the only species that suffer from rickets. It also occurs in most other vertebrates deprived of adequate sunshine. Before the cleanup of London's smog, it was observed in chimpanzees, lions, tigers, bears, deer, rabbits, ostriches, pigeons, and even lizards in the London Zoological Garden.

Now scientists know that rickets may be caused by a diet deficient in calcium and phosphate salts, or more commonly, by either the lack of vitamin D in the diet of humans (and mammals and birds) or inadequate exposure to sunlight.

That lack of sunlight could induce rickets in mammals on a vitamin D-deficient diet was demonstrated in the twenties, shortly after it was found that rickets could be cured by administering cod liver oil or by exposure to ultraviolet rays or sunlight. It was shown then that exposure to ultraviolet radiation of some foods containing oils with the complex organic alcohols known as sterols (but not cholesterol, often implicated in heart disease) resulted in the formation of curative substances. Vitamin D, as shown chemically in Figure 3:2, exists in several forms: The form obtained by irradiation on the skin is vitamin D_3, or cholecalciferol; it is made from a chemical precursor known as 7-dehydrocholesterol present in living skin cells. The ''creation'' of the vitamin from its precursor through a ''photochemical'' process involving UV-B radiation occurs in several steps, during which individual chemical bonds are altered by the absorption of ultraviolet radiation. It is thought that vitamin D_3 is the vitamin D characteristic of animals. In plant materials the vitamin produced by UV-B irradiation is vitamin D_2, or ergocalciferol, a ''photoproduct'' of UV-B irradiation of ergosterol common in some plant cells. The intermediate stages in the formation of vitamin D_2 are similar to

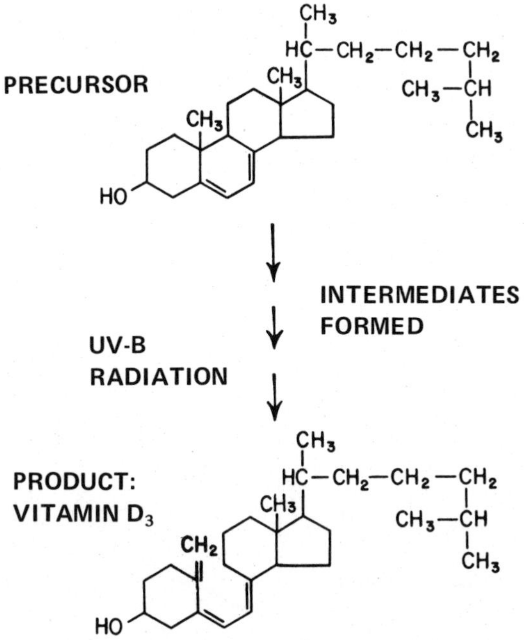

PRECURSOR

INTERMEDIATES FORMED

UV-B RADIATION

PRODUCT: VITAMIN D_3

FIGURE 3:2. Structural formulas of the precursor to vitamin D_3 and the product vitamin D_3. Between precursor and vitamin occur three steps for which UV-B absorbed by the skin of man supplies the energy. For details, see DeLuca, 1971.

those in the formation of vitamin D_3. The D vitamins have been named calciferols (cholecalciferol for vitamin D_3 and ergocalciferol for D_2) because of their calcifying action in bone.

Curiously, vitamin D in fish oils does not come from exposure of fish to sunlight, nor does it derive from fish eating invertebrate foods or algae containing vitamin D or its precursor. It is likely that fish oil vitamin D is formed by internal metabolic processes in fishes. The vitamin of fish oils is D_3, the same as the vitamin formed by UV-B irradiation in the living cells of human skin. Birds and mammals, apparently, obtain their vitamin D in the course of preening, the synthesis occurring in the oil from the sebaceous glands present on the surface of hair and feathers.

Vitamin D activity, as presented in Figure 3:3, involves mobilizing calcium, apparently by activating a transport system that takes calcium from the gut and places it into the bloodstream. Vitamin D_3 is oxidized in

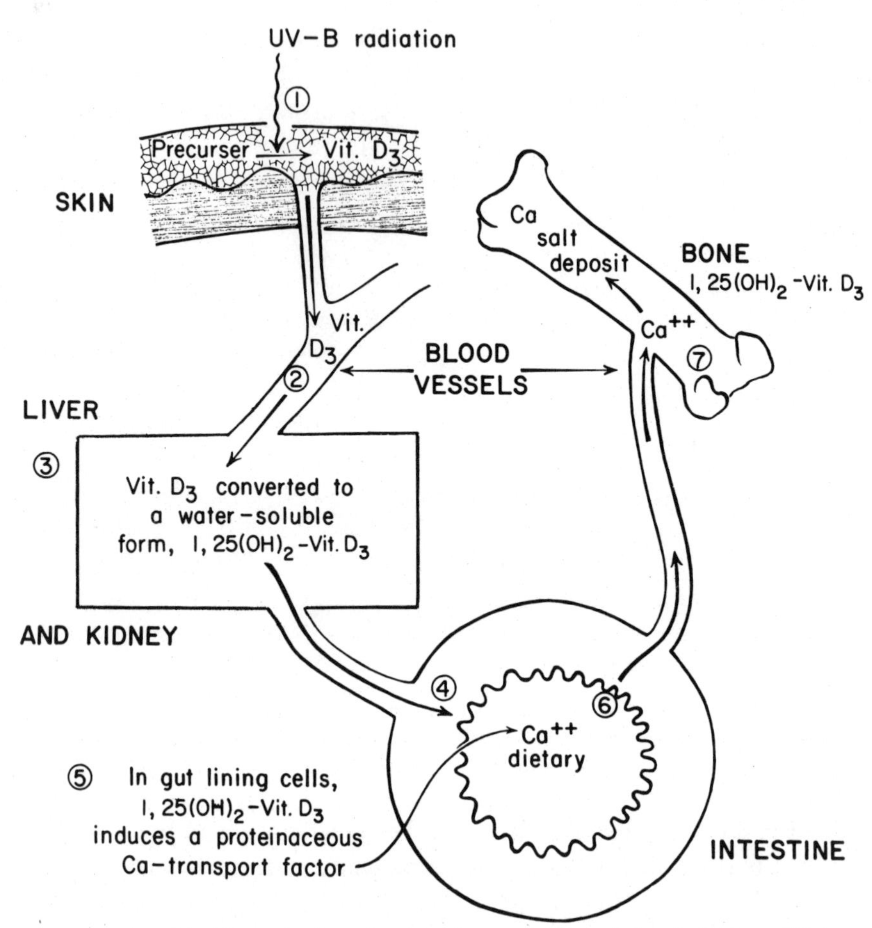

FIGURE 3:3. Possible mechanism of action of naturally formed vitamin D_3 in mobilizing salts for the maintenance of normal calcification of the bones. (Dietary vitamin D, chiefly D_2, comes from the gut to the liver by way of the circulation.) See text for details. For deposit of calcium and phosphate salts in the bone, the hormone of the parathyroid glands (small glands beside the thyroid) is required in addition to the water-soluble form of vitamin D (1,25-dihydroxychole-calciferol), which circulates in the blood.

two steps, first in the liver and then in the kidney to a form more water-soluble than vitamin D itself. It enters the cells of the intestinal lining and penetrates into the cellular nucleus, where it associates with the chromosome's genetic material, after which it triggers the formation of a messenger molecule with specific instructions to create a protein that carries calcium. The protein transports calcium through the cells of the gut to the bloodstream and ultimately to the bones. Note that vitamin D_2 acts in the same way.

Paradoxically, after long years of campaigning for a diet richer in vitamin D, in the United States at least, we may now be offering our children too much. Normal exposure to sunlight actually provides enough vitamin D to satisfy healthy bodily processes for most of us. When we supplement it with vitamin pills, enriched breakfast foods, milk, and other sources of vitamin D, we probably exceed our needs. Too much vitamin D may lead to calcification of soft tissues, such as the kidney, heart, aorta, and muscle. Hypercalcemia, excess calcium and phosphorus in infants, results from overdoses of vitamin D. If you supply vitamin D to a child at a dose one hundred times that required to cure rickets, calcium deposits will form in soft tissues. One thousand times that dose is fatal. Consequently, vitamin D enrichment of foods has been considerably reduced, and currently some physicians recommend control of vitamin D enrichment as well as reduced sale of vitamin D capsules. Surprisingly, because vitamin D mobilizes blood calcium, excess intake may even lead to bone decalcification. Muscular weakness, joint pains, and various other symptoms may follow.

Vitamin D may have influenced the evolution of skin coloration. The light skin of northern Europeans has selective advantage to these people in their winter climate, because even with little sunlight some antirachitic ultraviolet radiation penetrates lightly pigmented skin the better to convert the precursor to vitamin D. A child whose skin does not absorb enough UV-B radiation for vitamin D synthesis can develop rickets and, in severe cases, die. Light skin, however, is subject to more rapid degeneration and cancer induction than dark skin on overexposure to unlight.

Conversely, the dark skin of southern and tropical peoples allows a smaller fraction of antirachitic ultraviolet radiation to penetrate the skin, resulting in a lower rate of vitamin D synthesis, other conditions being the same. Fortunately, because of the greater number of sunlit days in the south, sufficient vitamin D is produced. The dark skin pigment also screens out the ultraviolet radiation causing sunburn, thereby greatly raising the threshold for skin damage during summer. Degeneration of the skin and cancer induction are thus rare for dark-skinned peoples.

When light- and dark-skinned people live in their native climates, each is well adapted to its habitat. If, however, light-skinned individuals migrate to sunlit southern climates, they show earlier and more prevalent skin damage and skin cancer. Dark-skinned people, on the other hand, who migrate north show greater incidence of rickets unless their diet receives supplements of vitamin D.

When vitamin D is produced by the skin under natural conditions, an excess is unlikely because the amount of precursor is limited. For this reason dark-skinned children are much less prone to overdoses of vitamin D, probably because their skin forms vitamin D photochemically less readily than light-skinned children.

For Additional Reading

Berven, H. "Physical Working Capacity of Healthy Children and the Effect of Ultraviolet Radiation and Vitamin D Supply." *Acta Pediatrica Supplement* 148 (1963): 3–16.

*DeLuca, H. F. "Vitamin D: A New Look at an Old Vitamin." *Nutrition Review* 29 (1971): 177–181.

———."Vitamin D: The Vitamin and the Hormone." *Federation Proceedings* 33 (1974): 2211–2219.

Fitzpatrick, T. B., Pathak, M. A., Harber, L. C., Sieji, M., and Kukita, A. eds. *Sunlight and Man.* Tokyo: Tokyo University Press, 1974, pp. 569–574.

Harm, W. "Biological Determination of the Germicidal Activity of Sunlight," *Radiation Research* 40 (1969): 63–69.

*Licht, S. "History of Ultraviolet Therapy." In *Therapeutic Electricity and Ultraviolet Radiation,* 2nd ed. edited by S. Licht. New Haven, Conn.: Elizabeth Licht, 1967, pp. 191–211.

*Loomis, W. F. "Rickets." *Scientific American* 223 (Dec., 1970): 76–91.

Seelig, M. S. "Are American Children Still Getting an Excess of Vitamin D?" *Clinical Pediatrics* 9 (1970): 380–382.

Urbach, F., ed. *The Biologic Effects of Ultraviolet Radiation.* Oxford: Pergamon Press, 1969. pp. 657–661 and 673–680.

Wurtman, R. I. "The Effect of Light on the Human Body." *Scientific American* 223 (July, 1975): 69–77.

4

How Ultraviolet Radiation Affects the Cell

When cells are exposed to sunlight, they absorb some ultraviolet radiation, chiefly in the UV-B range. The cell's nucleic acids and proteins take up most of the radiation, which often produces photochemical changes. After exposure, the altered molecules may affect one or more of the cell's functions. In complex multicellular animals like ourselves, UV-B radiation only penetrates, and is absorbed by, the cells near the surface; cells deep in tissues remain protected. UV-A radiation, however, goes into deeper tissue cells through the outer surface (epidermis) and some of it even penetrates the entire skin layer.

Radiation and Its Action

All wavelengths of radiation have certain properties in common. For example, they all travel as waves at the same velocity (3×10^{10} cm per second) and act as if they consisted of separate packets of energy known as quanta.* The energy in a quantum is inversely proportional to the wavelength. To be absorbed, a quantum must have energy equal to the difference in energy level between two excitation levels in a molecule or atom.

*A quantum is related to wavelength by the equation: $e = hc/\lambda$, where e is the energy in a quantum, h is Planck's constant, c is a constant—the velocity of the radiation—and λ is the wavelength of the radiation.

When a quantum of radiation is absorbed, the molecule becomes "activated" or "excited." The acquired energy may cause a photochemical change in the excited molecule, or it may be re-emitted as radiation of the same wavelength (resonance) or longer wavelength (fluorescence and phosphorescence), or it may be dissipated as heat. In very efficient photochemical reactions, where every radiation-absorbing molecule is transformed, the quantum efficiency is said to be one. If only a fraction of the excited molecules react, the quantum efficiency is less than one, but if the excited molecule produces excited products, a chain reaction may result in a quantum efficiency greater than one.

The effect of radiations of different wavelengths on entire organisms ultimately depends on the effect of the particular radiation on the cells of which the organism is composed and, in the cells, on the presence of absorbing molecules. Radiation must be absorbed to produce an effect on a molecule, since only absorbed energy can do the work necessary to promote a chemical change.

We see visible light because the pigment in the receptor cells in the retina of our eyes absorbs the light, and then notifies our brain through a series of electrical and chemical events. We know that visible light penetrates into deeper tissue cells in our bodies; most of us remember pressing a flashlight against our palm to see the red glow pass through our hand. In green plants, chlorophyll absorbs wavelengths at both ends of the visible spectrum and reflects the rest, which we see as green. Cells transparent to visible light appear colorless, and the light transmitted through them has no effect. Nucleic acids and proteins in living cells, in the absence of photosensitizers, are essentially transparent to visible light and UV-A radiation, but they absorb UV-B and UV-C radiation and are damaged by it. If we could see UV-C radiation, the chromosomes in cells would appear black to us. The photographs in Figure 4:1, taken with UV-C radiation, show chromosomes as either gray or black because the nucleic acids in the chromosomes absorb UV-C rays strongly.

The effect of a given quantum of absorbed light depends in part on the stability of the absorbing molecule. Thus, melanin, the protective pigment in our skin, is brownish-black because it absorbs most of the visible light. However, melanin, a stable molecule, is not altered perceptibly by visible light or even ultraviolet radiation and the absorbed energy is given off as heat. Melanin may also protect by reacting with fragments of molecules produced in the skin by UV-B radiation. The visual pigment

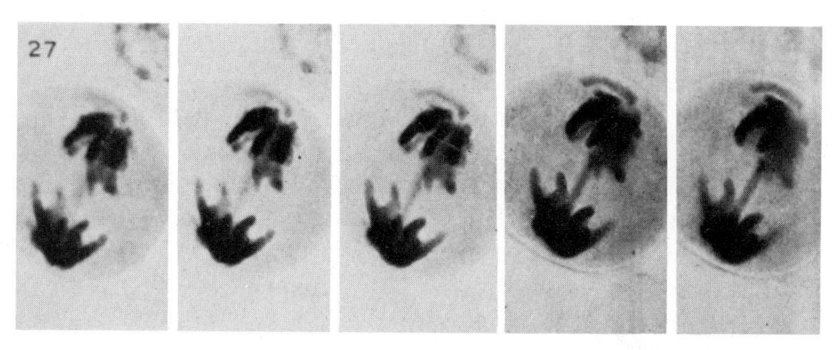

FIGURE 4:1. Chromosomes of a dividing cell appear gray or black in a photomicrograph taken with UV-C radiation because the DNA in the chromosomes absorbs so much of the incident radiation (living unstained sperm-forming cells of a grasshopper (*Melanoplus femur rubrum*) under quartz optics). The five figures are for the same cell at different focal levels, magnified 1200 times. In this phase of division (anaphase, secondary spermatocyte) the chromosomes condense at opposite poles of the cell. From Lucas and Stark, *Journal of Morphology* 52 (1931), p. 91, plate 3.

in our retina's rods and cones, on the other hand, is altered by absorption of visible light, and the molecular changes that occur are detectable by changes in absorption and breakdown products.

The effect of quanta of absorbed radiation from different parts of the spectrum upon cells depends in part on their energy content. Keep in mind that the shorter the wavelength, the greater the energy. For example, infrared quanta, which are small, alter the lower excitational levels of a molecule (rotational and lower vibrational states), but they cannot produce photochemical breakdown of the type induced by quanta of UV-B and UV-C radiation, which affect the higher excitational levels of a molecule (higher vibrational states and electronic excitation).

Clearly, how absorbed light affects a molecule depends on whether the quantum can raise the molecule to a higher vibrational or electronic excited state. It is, therefore, to be expected that a photochemical effect will not be influenced by the heat energy of the molecules at the time of light exposure, because heat affects lower energy levels in the molecule, below those involved in photochemical excitation. A rise of 10°C, which doubles the rate of a thermochemical reaction, thus has essentially no effect on a photochemical reaction.

Since the amount of a photochemical reaction induced by absorbed radiation depends on the number of molecules excited by light energy, it should not matter whether the molecules are exposed to low-light intensity for a long time, or high-light intensity for a short time, provided the product of the intensity by the time is the same. This is called the reciprocity law. This principle is familiar to photographers, and photochemical and photobiological reactions generally show agreement with this relationship. However, when rapid repair of radiation-induced damage occurs in cells, deviation from the reciprocity law is observed at low intensities of light when there is more time for repair to occur.

The chemical nature of molecules determines the absorption of nonionizing radiation, especially when atoms connected by alternate double bonds between them (carbon to carbon and carbon to nitrogen) and atoms in ring-shaped molecules are present, as shown in Figure 4:6. When different types of double bonds and rings are present, as in nucleic acids and proteins, absorption occurs at different wavelengths and to different degrees (see Figure 4:2). Consequently, absorption spectra are very important characteristics of molecules.

The Action Spectrum. The more a molecule absorbs radiation, the more effect radiation has on it. So, for example, in shortwave ultraviolet

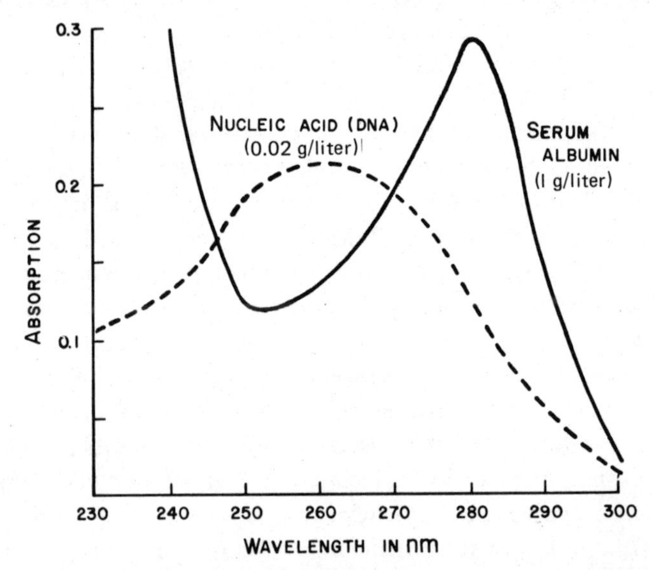

FIGURE 4:2. Absorption spectra of nucleic acid and a protein (serum albumin). The absorption is greatest at wavelength 260 nm for nucleic acid, at wavelength 280 nm for protein. The absorption by a very low concentration of nucleic acid (0.02 g per liter) is almost as great as that by 50 times as much protein (1 g per liter), indicating its greater absorbance, per unit weight, in this range of wavelengths.

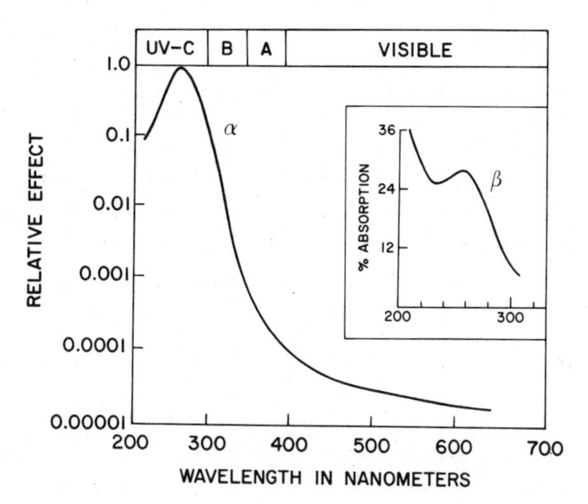

FIGURE 4:3. The bactericidal action spectrum showing the relative effectiveness of various ultraviolet and other wavelengths of radiation in killing bacteria is given by curve α. UV-C radiation is most effective and the effect of wavelength 260 nm is taken at 1.0 (most effective). UV-B and UV-A are less effective than UV-C radiation; however, even visible light at 400 nm (violet) kills bacteria, but it is only one ten-thousandth as effective as UV-C radiation at 260 nm. After Buttolph, in *Radiation Biology,* vol. 2, edited by A. Hollaender, McGraw-Hill, New York, 1955, p. 46. The relative effectiveness of various wavelengths of radiation for damaging various other types of cells is somewhat similar; for example, cells in our skin are damaged by doses of UV-A radiation only hundreds to thousands of times greater than UV-B radiation. Insert (curve β) shows percent of light at various ultraviolet wavelengths absorbed by a thin layer of bacteria. While maximal absorption occurs at 260 nm, the relative absorption at shorter wavelengths resembles absorption by proteins (see Figure 4:2). After Gates, *Journal of General Physiology* 19 (1930), p. 31.

radiation oxygen, the precursor, is converted to ozone, the product. Take the same number of quanta and deliver them at different wavelengths. How much photochemical conversion from precursor to product occurs depends on how many of the quanta are absorbed. The action spectrum obtained by plotting the amount of photochemical conversion from precursor to product along the vertical axis of a graph against the wavelength on the horizontal axis will correspond to the absorption spectrum at various wavelengths of some precursor molecule. The action spectrum, then, is just a measure of the relative efficiency of different wavelengths in converting precursor to product.

For scientists studying the effects of radiation on life, the action spectrum is a useful tool for tentatively identifying the precursor molecule, or receptor. The action spectrum for killing bacteria, called the bactericidal action spectrum, for example, corresponds to the absorption spectrum for nucleic acids as shown in Figure 4:3 (curve α). Researchers have found that both the bactericidal effect, and nucleic acid absorption (Figure 4:2), peak at 260 nm, and the entire curves parallel one another. The absorption spectrum for the bacteria also peaks at 260 nm, but at shorter wavelengths the curve parallels absorption by proteins as shown in Figure 4:3 (curve β). From this observation, it was concluded that bacteria are probably killed by nucleic acid damage and not by disrupted proteins. Since DNA (deoxyribonucleic acid) synthesis can be halted by lower doses of ultraviolet radiation than RNA (ribonucleic acid) or protein synthesis, investigators concluded that bacteria are killed by the effects of ultraviolet radiation on cellular DNA.

Biologists can identify the role of ultraviolet radiation in killing bacteria, but the action spectra for curative effects of ultraviolet rays in skin tuberculosis, in other skin diseases, and the presumed general beneficial "toning up" effects of ultraviolet radiation have not been determined. To be sure, it would be difficult to find out how ultraviolet radiation operates in these cases, because these studies would require high-intensity coverage over large areas, and repeated exposures for extended periods.

How UV-B Rays Kill Cells. The cell's response to the sun's ultraviolet radiation is the key to an understanding of the effects of sunlight

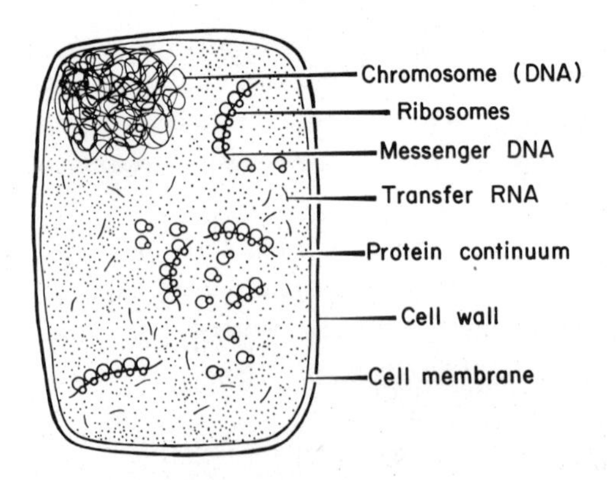

FIGURE 4:4. Diagram of a nonnucleated cell as exemplified by a bacterium. Note that the chromosome lies directly in the cytoplasm. Some bacterial cells move by hairlike structures called flagella (not shown). Blue-green algal cells are similar in structure to bacterial cells but have, in addition, membranous photosynthetic structures containing the green photosynthetic pigment chlorophyll.

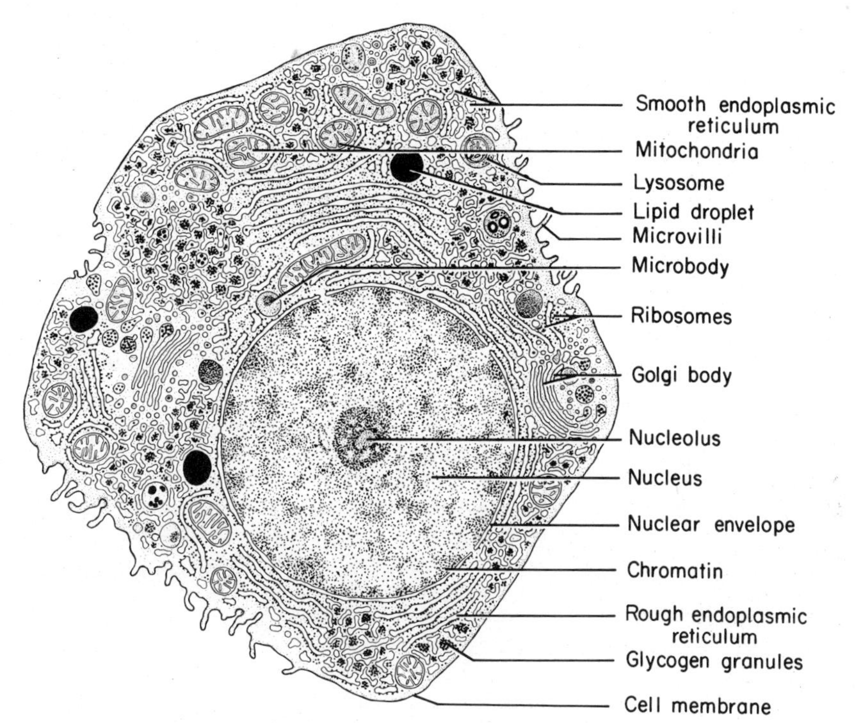

FIGURE 4:5. Diagram of a nucleated true cell, as exemplified by a cell from the liver (a hepatocyte). Between cell divisions the nucleus, enclosed in a double-membrane envelope, has diffuse genetic DNA called chromatin. The continuum of the cell between the organelles is proteinaceous as in a bacterium. Chromosomes generally appear during cell division (see Figure 4:1). From Lentz, *Cell Fine Structure,* Saunders, Philadelphia, 1971, p. 189.

on bacteria, plants, animals, and humans. A cell is a self-replicating unit of living substance enclosed in a cell membrane. As seen in Figure 4:4, cells of bacteria and blue-green algae are nonnucleated; the genetic material is present in a chromosome, which is a long, circular, tightly coiled molecule of DNA that lies directly in the cytoplasm, in contact with the cell membrane. Ribosomes composed of RNA and protein, messenger RNA, transfer RNA, and some of the enzymes that catalyze cell reactions are all involved in protein synthesis. Cell division requires normal replication of its DNA.

The cells of organisms other than bacteria and blue-green algae are more highly organized, as detailed in Figure 4:5. A nuclear envelope separates nucleus from cytoplasm and the chromosomes are usually multiple

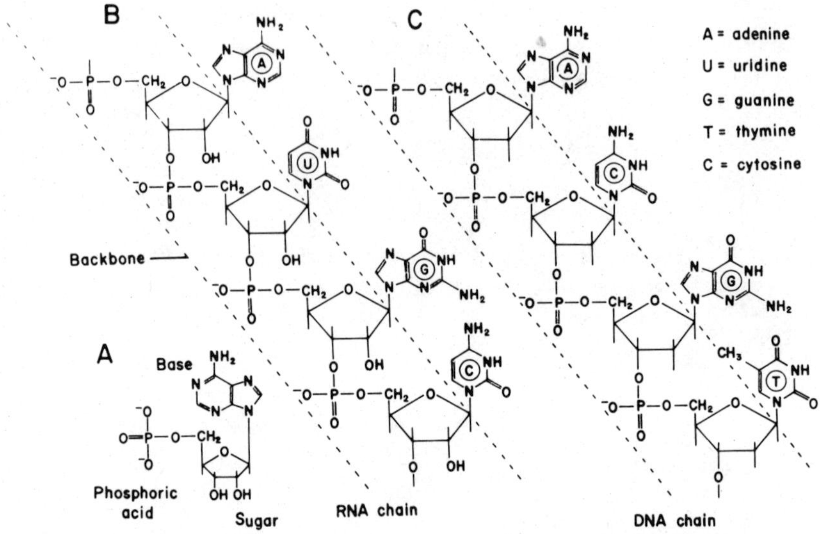

FIGURE 4:6. Nucleotides present in DNA and RNA strands. In the lower left-hand corner is shown a single nucleotide, like those linked by way of the sugar molecule in nucleic acids. Note the backbone of the nucleic acids composed of the phosphoric acid and the pentose sugars (ribose in RNA, deoxyribose in DNA) and the laterally extending bases. A stands for adenine, U for uracil, C for cytosine, G for guanine, and T for thymine. After Mazur and Harrow, *Textbook of Biochemistry,* 10th ed., Saunders, 1971, p. 158.

and much larger than the bacterial chromosome and, in addition to DNA and some RNA, contain protein; they are condensed and obvious only during cell division (see Figure 4:1). The cytoplasm is also compartmentalized with each part (the organelle) enclosed by membranes. Among the organelles is the endoplasmic reticulum, on which various molecules are synthesized, and many ribosomes may be attached to its surface. The mitochondria are the center of energy release, lysosomes package a variety of cellular enzymes, and so on.

The genetic material of the chromosome, the cell's encyclopedia of hereditary information, consists of DNA. DNA is composed of two antiparallel strands, each of which is a chain or polymer of nucleotide units. Each nucleotide (Figure 4:6A) in turn consists of three parts: an organic base, a phosphoric acid, and a five-carbon sugar (deoxyribose). The phosphoric acid and sugar form the backbone of the DNA strand with the attached bases protruding laterally as shown in Figure 4:6C. The two strands of the DNA molecule are held together by weak bonds (hydrogen bonds).

The uniqueness of each nucleotide depends on the four constituent organic bases, which in DNA are thymine, cytosine, adenine, and guanine. Thymine on one strand of DNA is always apposed to adenine on the other strand and cytosine is apposed to guanine, in accordance with a principle known as complementation, with the complementary pairs interlocked by hydrogen bonds, as shown in Figure 4:7.

Preceding cell division, the two strands of the DNA molecule separate from one another when DNA synthesis (called replication) is about to occur, each one acting as the template for synthesis of the new strand complementary to it, as Figure 4:7 also shows. An enzyme known as DNA polymerase adds a nucleotide to the nucleotide preceding it on the developing strand, forming a new strand in accordance with complemen-

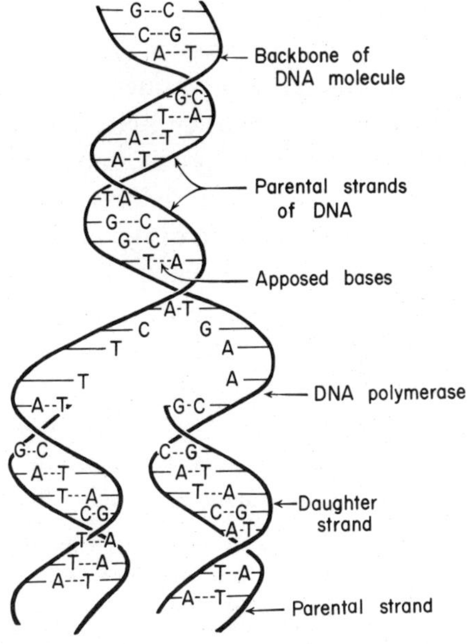

FIGURE 4:7. Diagram of replication of DNA. Note in the upper portion of the diagram the two-stranded nature of DNA and the hydrogen bonds (dashes) between complementary bases holding the two strands together. In some way, the DNA-polymerase (not shown) separates the two strands and synthesizes a new complementary strand on each strand, with the "old" or "parent" strands acting as templates. The DNA polymerase structure is not at present known, but it probably encloses the strands. Transcription of RNA occurs in a similar manner on a single strand of DNA, by action of an RNA-polymerase in the presence of the appropriate nucleotides typical of RNA and of phosphate compounds as energy sources. (RNA is single-stranded though in parts of its length it may hydrogen-bond to itself forming clover leaf configurations.)

tation. An adenine-containing nucleotide is added in correspondence with a thymine-containing nucleotide on the template strand, while a cytosine-containing nucleotide is added in correspondence with a guanine-containing strand on the template strand, and so forth. In this way, each newly synthesized strand is complementary to the template strand, but the new double-stranded DNA molecule as a whole is a duplicate of the parental DNA, barring a mistake in replication.

RNA is much like DNA in structure, except that the base uracil replaces thymine and a different sugar molecule is present, as seen in Figure 4:6B. RNA synthesis (transcription of the message from DNA) occurs along a single strand of a DNA molecule under guidance of the enzyme RNA polymerase.

The RNA molecule is complementary to the DNA molecule (with uracil in place of thymine, as noted above). Thus, any code of information represented by the sequence of bases on DNA is repeated in complementary form on RNA transcribed. Transcribed RNA varies in size: Ribosomal RNA is fairly large, transfer RNA is small, while messenger RNA varies with the size of the protein molecule for which it has information.

For protein synthesis (translation of the message from the DNA), messenger RNA attaches at a complementary site on the ribosome, as illustrated in Figure 4:8, and to it, carrying activated amino acids, come transfer RNA molecules that attach at complementary sites on the messenger RNA. In the presence of required enzymes, each amino acid is linked to the preceding amino acid, thereby generating a chain of amino acids characterizing a protein molecule, as Figure 4:8 shows.

Direction as to what amino acid to attach where comes from the code transcribed from DNA to messenger RNA. The "words" of the genetic code for the appropriate amino acids (of some twenty amino acids possible) consists of sequences of three bases in the DNA. Thus, AAA in DNA, transcribed to UUU in messenger RNA where U takes the place of T (but back to AAA on transfer RNA) means the amino acid phenylalanine, GAG in DNA, which transcribes to CUC in messenger RNA (but back to GAG on transfer RNA), means the amino acid leucine, and so on. In this simple genetic code language, which consists of four kinds of bases combined in the various possible three-base permutations, all the information required to form the multitude of proteins

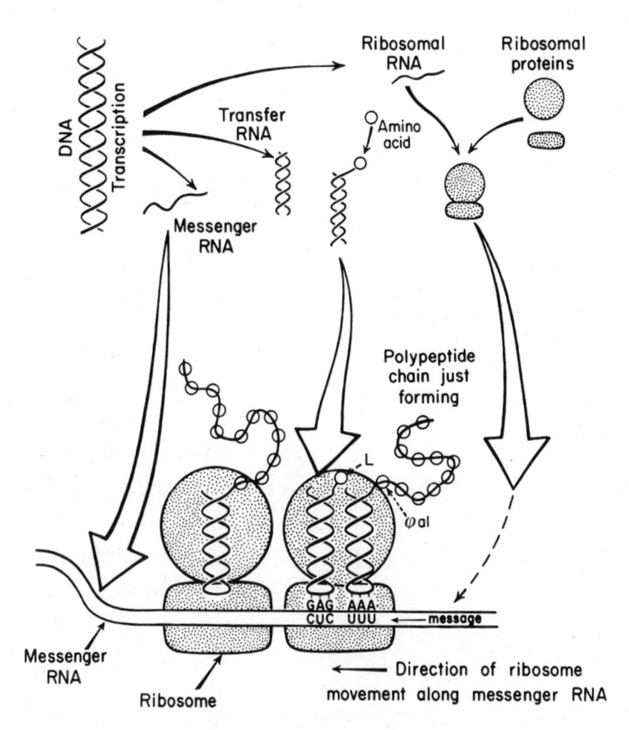

FIGURE 4:8. Diagram of protein synthesis on a ribosome. The information for the protein to be synthesized comes from the DNA on which the various types of RNA are transcribed. tRNA refers to transfer RNA, mRNA to messenger RNA, and rRNA to ribosomal RNA. rRNA combines with ribosomal proteins to form an active ribosome. The ribosomes combine with a messenger RNA molecule to which amino acids are carried by tRNA. Activation of amino acids is accomplished with the breakdown of a high-energy phosphate, adenosine triphosphate (ATP), to adenosine monophosphate (AMP) and guanosine triphosphate (GTP) to guanosine monophosphate (GMP). The arrows point toward enlargements of the units involved (for better visualization).

necessary to life's processes can be represented. "Breaking" the genetic code was one of the great achievements of our era.

The energy to run the assembly lines for synthesis of DNA, RNA, and protein is supplied by "high-energy" phosphate compounds that, like fuels, release energy when they decompose. Such phosphate compounds "activate" the reacting molecules.

Biologists have found that ultraviolet radiation damages cells by interfering with macromolecular (DNA, RNA, and protein) syntheses. DNA synthesis is the prime target, since DNA is not only necessary for all other cell syntheses, but it is present in least amount (usually in duplicate, sometimes only singly) whereas RNA and protein are present in multiple. Consequently, DNA is affected by the smallest doses. Cellular DNA, as summarized in Figure 4:9, can be altered in any one of a number of ways by ultraviolet radiation: (1) by adding water to thymine or cytosine bases (or what is called hydration); (2) by linking thymine to thymine (T-T), cytosine to cytosine (C-C), or thymine to cytosine (T-C) on the same strand of DNA (dimer formation); (3) by linking these bases across complementary DNA strands (not shown in figure); (4) by breaking phosphate bonds in the backbone of a DNA molecule; (5) by linking some protein to the DNA; and (6) by denaturation, the breaking of hydrogen bonds. Of these, the formation of dimers between thymine bases in DNA is usually the most frequent (see Figure 4:10). When DNA is extracted from irradiated bacteria, it shows the presence of base dimers as well as some of the other types of damage. With increasing doses, the linkages multiply and DNA replication falters. Reproduction is either seriously delayed or prevented in an increasing proportion of bacteria.

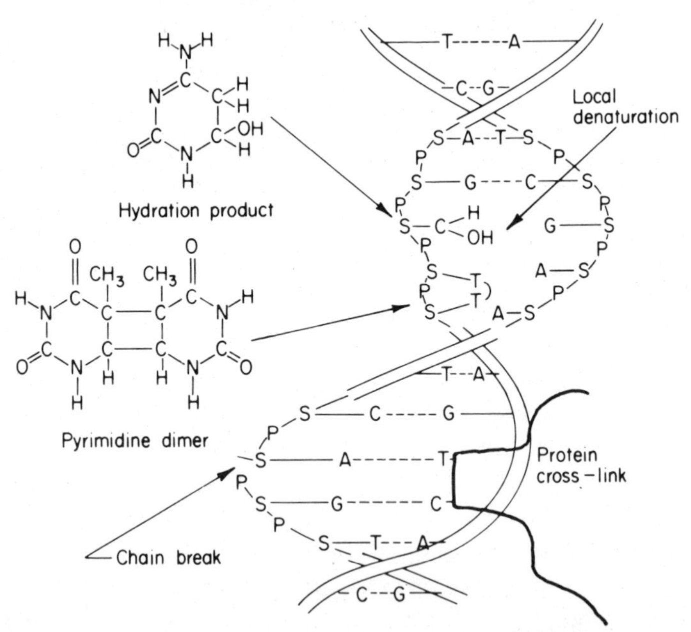

FIGURE 4:9. Various possible alterations in DNA extracted from UV-B- or UV-C-radiation treated cells. From Smith and Hanawalt, 1969, after Deering, *Scientific American* 207, 1962, p. 135.

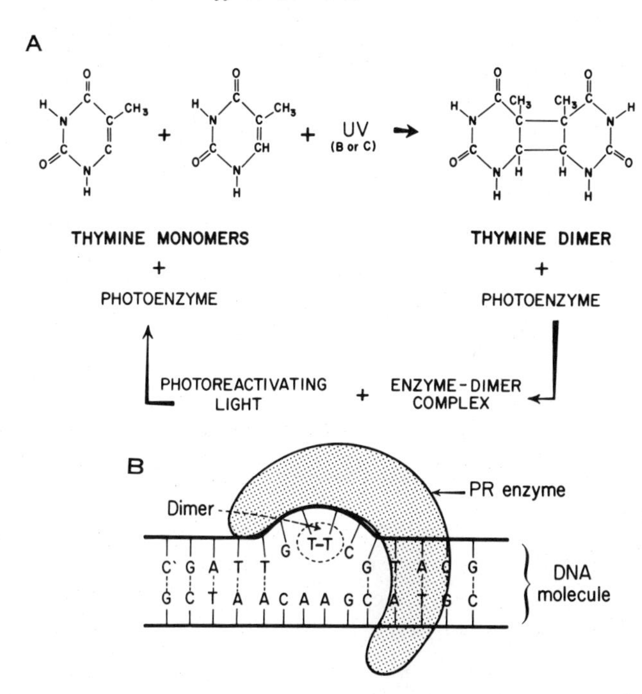

FIGURE 4:10. (A). Thymine dimer formation between neighboring thymine residues in DNA after exposure to UV-B or UV-C radiation; similar dimers are formed in smaller numbers between other neighboring pyrimidine residues in DNA. (B). Diagrammatic representation of possible structural distortion of DNA as a result of dimer formation and its removal by photoreactivation. The DNA-polymerase would probably be unable to proceed with DNA synthesis around such a block during normal replication. However, the distortion specifically attracts the photoreactivating enzyme (shown as a shaded commalike structure), which on absorption of light splits the thymine dimer (monomerization), restoring the DNA to its original state. After Hanawalt, *Endeavour* 31, 1972, p. 84.

In RNA, changes similar to those that occur in DNA probably exist also, but because there are many more molecules of each kind of RNA than DNA in the cell, many more molecules must first be altered before the effect of irradiation interferes with cell reproduction. Similarly, proteins may also be denatured during irradiation, but because large numbers of each kind of protein molecule are present in the cell, these effects do not interfere with reproduction until a large number of molecules have been denatured. Furthermore, as long as DNA is intact, more RNA and protein can be synthesized, replacing damaged RNA or protein.

Exposure to ultraviolet radiation blocks DNA replication—at least temporarily—because the three-dimensional (or steric) changes interfere with the enzyme DNA polymerase, which polymerizes the nucleotides to form new strands of DNA. The most common DNA damage following ultraviolet irradiation appears to be dimer base formation, and the most usual dimers are thymine to thymine (Figure 4:10).

UV-B rays affect single-celled animals and plants unprotected by pigment, eggs and sperm issued into environmental water, and cells in small colorless many-celled animals, in developmental stages of various aquatic invertebrates and fishes, and in unpigmented skin of animals and epidermis of plants. While details are lacking, the action spectra for killing various types of cells by ultraviolet radiation usually suggest damage to DNA, preventing replication.

We should not, in our attention to DNA, forget that synthesis of other macromolecules—RNA and proteins—may also be reduced or stopped by exposure to ultraviolet radiation, but larger doses are required. For example, an action spectrum resembling absorption by protein is obtained in inhibition of motility in ciliate protozoans. It is presumed that proteins (enzymic or structural?) in the cilia are altered by the radiation. Recovery from such injury in ciliates is more rapid than recovery from ultraviolet damage to DNA.

How Do Cells and Organisms Protect Themselves?

Motile single cells, small multicellular animals, developmental stages of larger animals, and motile stages of lower plants composed of a few cells can be damaged by sunlight ultraviolet radiation, but fortunately built-in behavioral responses enable many of them to move out of the sun and live beyond reach of destructive rays. Many protozoans randomly move into protective hiding places and avoid injury. The phytoplankton composed of floating microscopic plants and the major crop in fresh and sea water grow some meters below the surface of the water, avoiding direct attack by the sun. No one knows whether this is a specific reaction to UV-B radiation; but we know that these cells do respond to radiation.

The mass of marine zooplankton—made up of single-celled animals and larval stages of multicellular animals—found below the phytoplankton in the sea during the day, feed as they swim upward through the phytoplankton with the approach of darkness and return below the phytoplankton again at dawn. Marine biologists do not yet know whether UV-B radiation has anything to do with their apparent light-avoiding movements and their distribution.

Because of its shallow penetration, UV-B radiation affects only the skin of animals and humans, especially if pigments and other sun screens are absent. Many animals avoid noonday sunlight. Those who have watched range animals have witnessed them seeking shade on hot days. Wild animals rarely stay in direct sunlight either, especially on warm days. Many animals are more active at dusk and night and seek the shadows by day. The heat, as well as the greater protection from predators afforded by a forest cover, probably causes most animals to avoid the sun. People also avoid the sun on hot days because of the heat load; the darker their skin, the more easily they become overheated.

In the desert, animals lead an active night life and dawn usually signals their return to moist underground burrows or the shade of stones. The cue to avoid sunlight may also be the rapid loss of water. Above ground, moisture is lost not only from the surface of the body but also by breathing.

However, on cold days cold-blooded animals such as insects and reptiles are cold-immobilized and bask in the sun to warm themselves to activity. Even birds and mammals, though warm-blooded, may bask in the sun on cold days.

Seed plants cannot move out of sunlight but some, such as compass plants, turn their leaves in such a way as to minimize exposure. Other plants, unable to turn their leaves, may move chloroplasts in individual cells to shade one another, perhaps sacrificing a few to achieve minimal exposure to the mass of chloroplasts in the cell. In neither case is it clear that light, rather than heat, induces the movements, though chloroplast alignment in aquatic plants would appear to be an effect of light because the heat of sunlight would be rather quickly dissipated in water.

Seeds of shade plant species—ill adapted to growth in direct sunlight—germinating in areas exposed to the sun simply do not make it to maturity. Only seedlings of species with the appropriate adaptation for existence in the sun survive. Shade plants moved into sunlit areas, even if provided with adequate water, will be damaged by the sun.

Sunscreens. Some cells carry protective pigments to absorb sunlight and minimize its damage. While not common, pigments may even be present in single-celled forms of life. For example, the black pigment in the ciliate *Fabrea* enables it to withstand all-day exposure in salt pools.

Multinucleate slime molds have various pigments and it is possible to see brown and black spores on fungi. In higher organisms, pigments are found in the cells on the outer layer of skin. Animals are usually pigmented and people native to the sun-drenched regions of the earth are also usually deeply pigmented. Even sun-tolerant plants often have a pigmented epidermis.

Albinos or partial albino individuals occur in nature in various species of animals and plants and in all races of man. It is their misfortune to be highly sensitive to UV-B radiation in sunlight. In the laboratory, scientists produce albino animals, but the wild type in each species is pigmented. In nature those species in which all individuals are albino occur almost exclusively in caves or in burrows. Curiously, deep-sea animals are generally pigmented, even though little if any light reaches them.

Skin pigment may also be increased on exposure to sunlight, as in the tanning of human skin. The coloration in the epidermal cells of many plants may also be deepened after half a day's sunlight exposure. In the sun, these plants produce flavones, pigments like those responsible for autumn colors in leaves.

Animals may be coated with hair, feathers, scales, hides, or shells—all impervious to sunlight. Even the skin of white-haired animals is protected from sunlight except where the unpigmented skin lacks hair—around the mouth, nose, eyes, etc. Some plants develop a thick cuticle and waxy coating, while others have hairy outgrowths of epidermal cells. Probably all these devices protect against heat and evaporation, but they also serve to protect against intense sunlight by preventing UV-B radiation from reaching the sensitive components of the epidermal cells.

Photoreactivation (Enzymatic Repair). To a considerable extent the damage inflicted by exposure to UV-B and UV-C rays can be reversed by simultaneous or subsequent cellular exposure to the blue-violet or UV-A radiation. This process, known as photoreactivation, occurs in all cells with photoenzymes. In studies with extracted DNA, this enzyme attaches to the base dimers (mainly thymine) of the UV radiation-treated DNA (see Figure 4:10). Neither altered DNA nor photoenzyme alone absorbs radiation in the photoreactivating part of the spectrum, but the complex of the two apparently can, resulting in the splitting (monomerization) of the dimer. This returns DNA to its original condition. Because the photoenzyme attaches only to dimers in DNA, it fails to reverse ultraviolet radiation-induced damage of other types to DNA or to RNA or to proteins. Since photoreactivation occurs to the extent of about 80 percent and only dimers are reversed, photobiologists have concluded that the major change in DNA induced by ultraviolet radiation in active cells appears to be production of mostly thymine base dimers.

Photoenzymes occur in cells of a wide variety of plants and animals and in bacteria, although they are lacking in some types of bacteria and, curiously enough, almost entirely absent from cells of placental mammals, a deficiency for which there is at present no explanation. Recently, however, a photoenzyme has been discovered in white blood cells and skin fibroblasts of placental mammals.

Photoenzymes have also been found in soil bacteria, in bacteria of the human intestinal tract, and in the deep-lying cells of multicellular animals and plants, protected by the overlying cells—all of which seldom if ever see the light of day. It is likely that the photoreactivating mechanism in the basic genetic pattern of cells has been retained because of its critical importance in the early evolution of life; or, alternatively, because it may have a function not involving light that is yet to be revealed.

Note that photoreactivation is never complete. Some damage to DNA never undergoes repair. Its backbone may become irreparably damaged, or linkage between DNA and protein may occur, or possibly vital proteins may be harmed.

Dark Repair. One of the innate properties of the living cell is its continual repair of damage to DNA by means of dark repair. In ultraviolet radiation-damaged cells of animals, including placental mammals, it occurs either in complete darkness or in light. The mechanisms of dark repair are quite different from photoreactivation, although dimers may be repaired by either method. There are two general classes of dark repair: excision and postreplication.

Excision Repair. In this class of dark repair, often called cut-and-patch mechanism, a piece of the DNA, consisting of several nucleotides including the dimer, is excised under the influence of an enzyme. The excised portion of the strand is then replicated under the guidance of DNA polymerase from information given by the complementary DNA strand. The new piece is bound to the loose end of the uninjured portion of the DNA by the action of a binding enzyme (see Figure 4:11). Ultimately, DNA is restored to its original form. In life, excision repair

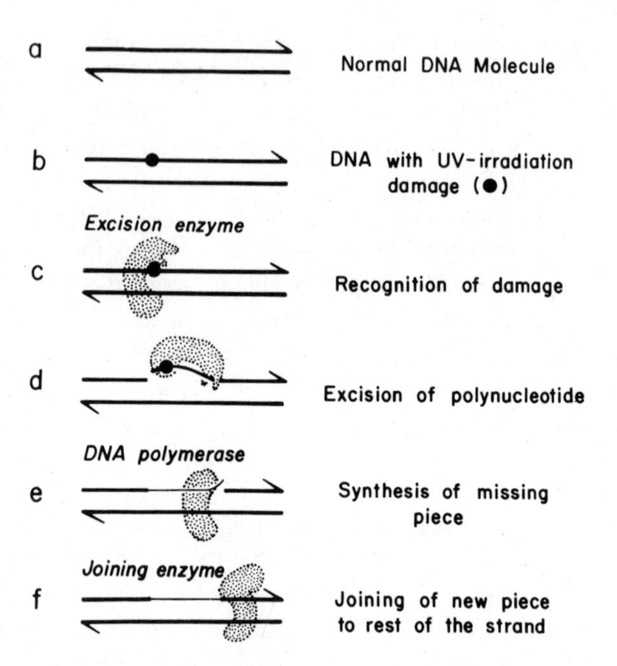

FIGURE 4:11. Possible mechanism of excision repair by "cut and patch" of UV-radiation-induced damage in a DNA molecule. DNA shown in a simplified model with anti-parallel strands (a) is injured by UV-B or UV-C radiation as shown by the black dot (b). When the injury is recognized (c), the excision enzyme snips it out as a polynucleotide (d). It is then replaced by a piece of DNA synthesized under action of the DNA-polymerase using the information on the complementary strand (e). The new piece is then attached (f) by the joining enzyme (ligase). After Hanawalt, *Endeavour* 31, 1972, p. 84. The differing shapes of the enzymes are for diagrammatic purposes only; their real shapes are not known.

seems to be widespread, although it has been studied primarily in bacteria and mammalian cells. Little evidence for the cut-and-patch approach has been found in plant cells; however, only a few studies have been made with these cells. Because it is so widespread, excision repair may be primarily used to correct small errors in routine replication known to occur with greater frequency than those that show up as mutations. Excision repair has recently been demonstrated in wild carrot cells. The repair system is overwhelmed by large radiation doses. It is likely that previously reported negative results on plant cells did not provide appropriate conditions for demonstration of excision repair.

Postreplication Gap Repair. This second class of dark repair occurs after replication has proceeded to either side of a damaged portion of a

DNA strand, leaving gaps. Repair then follows, after replication of the undamaged part, essentially by filling the gap(s) with the synthesis by DNA polymerase using the information on the complementary undamaged strand. This is followed by the insertion of the newly synthesized piece (or pieces) by a binding enzyme (Figure 4:12). While the facts are clear, the way in which this type of dark repair works is still unresolved.

Mostly, postreplication repair has been studied in bacteria, but considerable work on cells in tissue culture demonstrates that it also occurs in mammalian cells. There is a suggestion of such repair in a primitive plant cell, but at present no experiments have been done with cells of higher plants.

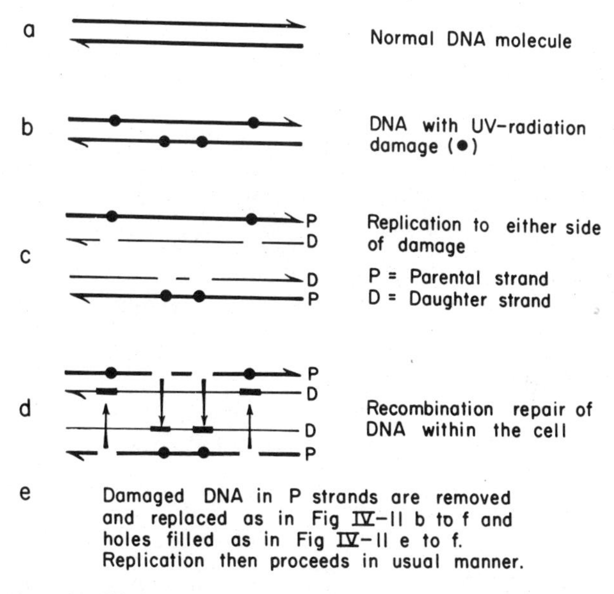

FIGURE 4:12. A simplified model for postreplication repair of UV-damaged DNA molecules (shown as antiparallel lines with arrows): (a) normal DNA molecule; (b) dots indicate damage in DNA; (c) DNA synthesis (replication) under direction of DNA polymerase (not shown) proceeds on either side of the damage in the parental strands, leaving gaps in the daughter strands. After replication (c) the missing pieces are added to the incomplete DNA daughter strands (D) by recombination with corresponding uninjured pieces of the parental DNA strands (P), within the cell. The remaining pieces of the parental DNA strands (d) may be repaired by enzymatic excision by the DNA polymerase of damaged regions in the DNA and insertion of polynucleotides in these gaps and in those resulting from recombination repairs, in the same manner as in Figure 7:11. It is possible that in some cases the gaps are repaired without recombination. After Smith, in *Sunlight and Man,* edited by Fitzpatrick *et al.,* University of Tokyo Press, Tokyo, 1974, p. 72.

Both types of dark repair can be done in the light as well as in the dark, but in the light photoreactivation probably occurs more rapidly, and thymine dimers may be split and native DNA reconstituted, even before the dark repair enzymes involved have had much opportunity to operate. However, damages to DNA other than pyrimidine dimers such as single strand breaks (Figure 4:9, lower left, chain break) are also subject to dark repair, lending dark repair the advantage of generality.

Damage to DNA produced by other agents such as x-rays and chemicals (especially carcinogens) or by errors in replication is also repaired by dark-repair mechanisms in cells.

Mutation

Mutation is an alteration of genetic material, a change in the genes found in a chromosome. Since DNA is the genetic material of cells, mutations are caused by changes in DNA. A gene expresses itself by signaling, through messenger RNA, the synthesis of a specific protein needed for some process in the cell. Even a small change in a protein, such as the substitution of a single amino acid, may change the specificity of the protein. For example, sickle cell anemia, a disease common among blacks, results from the replacement of only one amino acid in hemoglobin, the oxygen-carrying protein of the red blood cell. As a result, red blood cells, when kept at low oxygen, change from the characteristic disk of normal cells to a sickle shape. Remember that an amino-acid residue in a protein is coded by a set of three bases in three successive nucleotides in the DNA; a mutation is a change in that code for some amino acid. The change may result from the alteration of one of the three bases by ultraviolet radiation, or from the loss of one of the three bases, or from the addition of another base in the sequence—in all cases causing at least one three-base "word" to be "misspelled." This, in turn, means at least one amino acid in the sequence of amino acids in the protein coded by the altered gene may not be the normal one. The protein produced is, therefore, likely to be inactive or active in a different manner.

Mutation may be spontaneous or induced. A spontaneous mutation is the result of a mistake occurring during DNA replication with no observable outside disturbance of the process and with no correction by some cellular repair mechanism. An induced mutation is one that follows exposure to some disturbing environmental factor, such as ultraviolet radiation, which has been shown to accelerate the rate of mutation. To induce a mutation by ultraviolet radiation, a dose that kills the vast majority of the cell population must be used. This probably means that all of the cells have been damaged, but that repair mechanisms are active. It is assumed that, inasmuch as most of the cells do not recover, the repair mechanisms in most of the cells have been overwhelmed. In the few cells that survive, mutations occur at a higher rate than in a comparable population of control cells that have not been treated by ultraviolet radiation.

Why Repair?

When we compare the sensitivity to ultraviolet radiation of wild-type cells (the sort one would find in nature) with that of mutant cells that lack one or more of the repair mechanisms, we can easily understand the significance of repair mechanisms. Let us take, for example, a dose of UV-B or UV-C radiation that kills only a small fraction of the wild-type cells of the colon bacillus (*Escherichia coli*) placed in the dark immediately after exposure. Examination reveals that it has killed a large fraction of mutant cells lacking either the excision-repair mechanism or the postreplication repair mechanism. A double mutant, lacking both excision and postreplication repair, is vastly more sensitive, and most cells of this type are killed by a dose so small that it would have an almost undetectable effect on the wild type as evident in Figure 4:13. Double mutants, lacking both dark-repair mechanisms but having a photoenzyme, although exceedingly sensitivie to UV-B and UV-C radiation when placed in the dark immediately after exposure, may show a high degree of recovery if immediately placed in photoreactivating light. If the photoenzyme is also lacking in a mutant without dark repair, that mutant would not survive.

Animal cells throughout the animal kingdom, including the marsupial mammals such as the opossum, possess the photoenzyme, but cells of placental mammals lack it and show no photoreactivation, with the possible exception of certain white blood cells and fibroblasts as noted earlier. Injury to placental mammals from ultraviolet irradiation must therefore be repaired by dark-repair mechanisms in cells lacking the

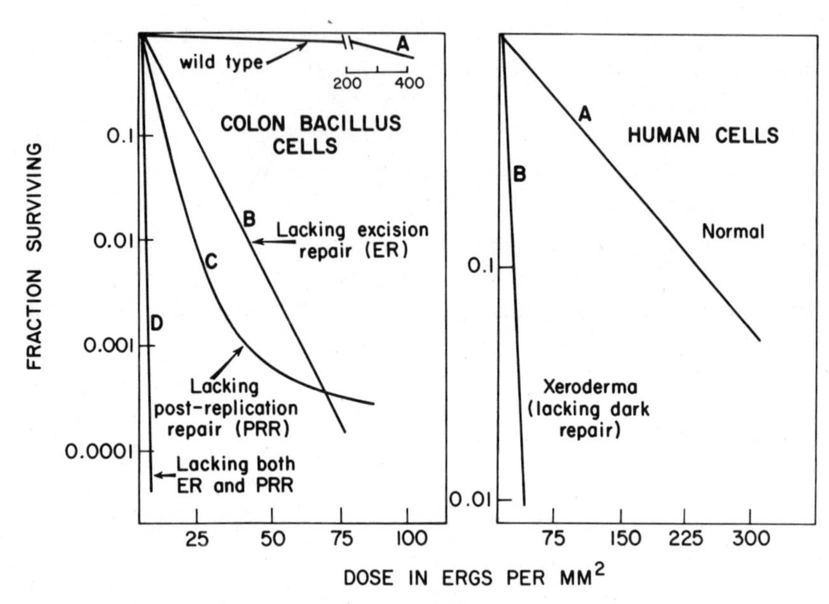

FIGURE 4:13. Left side: Comparative susceptibility to UV-C radiation of wild type and mutant colon bacteria lacking one or the other repair mechanism or both. The wild type is highly resistant to the radiation as shown by high percent survival even after a relatively large dose of radiation (400 ergs per mm^2), while the mutants lacking either recombination repair or excision repair show less than 0.01 percent survival. A double mutant, lacking both excision and recombination repair, is reduced to 0.01 percent survival by only a few ergs per mm^2 of the UV-C radiation. After Howard-Flanders and Boyce, *Radiation Research Supplement* 6, 1966, p. 156. Repair in various strains of the colon bacterium occurs to different degrees, not always to the maximal extent shown here; this indicates that a number of genes govern the degree of repair. Right side: Susceptibility to UV-C radiation (germicidal lamp) of skin cells from a xeroderma pigmentosum victim compared to those of a normal individual. After Takebe, Furuyama, Miki, and Kondo, in *Sunlight and Man,* edited by Fitzpatrick *et al.,* Tokyo University Press, Tokyo, 1974, p. 109. At least five genes are involved in dark repair in xeroderma with perhaps a graded series of resistance to ultraviolet radiation, but the details have not been as well worked out as in the bacteria because of the paucity of material.

photoenzyme. Both excision repair and postreplication repair have been demonstrated in some mammalian cells. When either of these is lacking, the cells show extraordinary sensitivity to sunlight. For example, the skin of a person with the disease known as xeroderma pigmentosum, lacking either excision repair or postreplication repair, is extensively damaged by a brief exposure to sunlight, as shown in Figure 4:13. Dark repair occurs in most animal cells tested but the mechanism has not received much attention, though excision repair has been demonstrated in a number of them.

The sun's ultraviolet radiation is probably harmful to all cells, but cells have the capacity to repair much of the damage inflicted. Continuation of life, including human life, obviously depends on the operation of repair mechanisms that have evolved hand in glove with the exposure of living things to damaging ultraviolet radiation in sunlight. Under the sun's ultraviolet rays, life leads a precarious existence, a delicate balance between damage and repair.

When we consider evolution, dark repair has an even broader significance. Since a wide variety of damage produced spontaneously or induced by environmental influences can be repaired in this way, the stability of genes is vastly increased, thus protecting the bank of genetic information passed on from one generation to another.

Most mutations result in the loss of some function or the decrease in production of some needed substance, and hence survival of the species of organism is at a disadvantage if the mutation occurs widely enough. Some mutations, on the other hand, may offer improvements in function and serve as material for selection.

For Additional Reading

Caldwell, M. M. "Solar UV Radiation and the Growth and Development of Higher Plants." In *Photophysiology,* vol. VI, edited by A. C. Giese. New York: Academic Press, 1971, pp. 131–177.

*Clayton, R. K. *Light and Living Matter,* vols. 1 and 2. New York: McGraw-Hill Book Co., 1970–1971.

Cook, J. S. "Photoreactivation in Animal Cells." In *Photophysiology* vol. 5, edited by A. C. Giese. New York: Academic Press, 1970, pp. 191–233.

Giese, A. C. "Effects of Ultraviolet Radiations on Some Activities of Animal Cells" In *The Biologic Effects of Ultraviolet Radiation,* edited by F. Urbach. Oxford: Pergamon Press, 1969, pp. 61–82.

Grossman, L. "Enzyme Involved in the Repair of DNA." *Advances in Radiation Biology* 4 (1974): 77–129.

Halldal, P. and Taube, O. "Ultraviolet Action and Photoreactivation in Algae." In *Photophysiology,* vol. 7, edited by A. C. Giese. New York: Academic Press, 1972, pp. 163–188.

*Hanawalt, P. C. and Haynes, R. H. "The Repair of DNA." *Scientific American* 216 (Feb., 1967): 36–43.

*Hanawalt, P. C. and Setlow, R., eds. *Molecular Mechanisms for the Repair of DNA.* New York: Plenum Publishing Corp, 1975.

*Haynes, R.H., and Hanawalt, P. C. *The Chemical Basis of Life.* San Francisco: Freeman, 1968.

Howland, G. P. "Repair of Ultraviolet-Induced Pyrimidine Dimers in the DNA of Wild Carrot Protoplasts." *Nature* 254 (1975): 160–161.

Jagger, J. *Introduction to Research in Ultraviolet Photobiology.* Englewood Cliffs, N.J.: Prentice Hall, 1967.

Menon, I. A., Gan, E. V., Lam, K. M., and Haberman, H. F. "Electron Transfer Properties of Melanins and Their Possible Role in Protection against UV Radiation." In Abstracts, 3rd Annual Meeting American Society for Photobiology, 1975, p. 115.

Rupert, C. S. "Photoreactivation of Ultraviolet Damage." In *Photophysiology,* vol. 2, edited by A. C. Giese. New York: Academic Press, 1964, pp. 283–327.

Setlow, R. B. "Action Spectroscopy" In *Advances in Biological and Medical Physics,* Vol. 5, edited by J. H. Lawrence and C. A. Tobias. New York: Academic Press, 1957, pp. 37–74.

Smith, K. C. "The Roles of Genetic Recombination and DNA Polymerase in the Repair of Damaged DNA." In *Photophysiology* vol. 6, edited by A. C. Giese. New York: Academic Press, 1971, pp. 209–278.

———. "Molecular Changes in Nucleic Acids Produced by Ultraviolet and Visible Radiation" and "The Cellular Repair of Radiation Damage." *Sunlight and Man,* edited by T. B. Fitzpatrick *et al.* Tokyo: Tokyo University Press, 1974, pp. 57–66 and 67–77.

———. (ed.) *Protein and Other Adducts to DNA: Their Significance to Aging, Carcinogenesis and Radiation Biology.* New York: Plenum Publishing Corp., 1976 (in press).

———. (ed.) *The Science of Photobiology* (a textbook). New York: Plenum Publishing Corp. 1976 (in press).

Sutherland, B. M. "Photoreactivation in Animal Cells." *Life Science* 16 (1975): 1–6.

*Wald, G. "Radiation and Life." In *Recent Progress in Photobiology,* edited by E. J. Bowen. Oxford: Blackwell, 1965, pp. 333–350.

Wolff, S. "Chromosome Aberrations Induced by Ultraviolet Radiation." In *Photophysiology* vol. 7, edited by A. C. Giese. New York: Academic Press, 1972, pp. 189–205.

5

Sunburn

Ancient legend tells us that when the sun chariot came too close to the earth it scorched life upon it. And today we know that too much sun not only burns our skin but also the surface of many animals and plants. Sunburn results chiefly from cumulative injury to cells in the exposed skin. Interaction between injured cells and other parts of the skin is a controversial area of scientific study and few clear results are reported. This chapter will consider the cellular events.

When the sun first strikes human skin, it causes a reddening or what is clinically called erythema (the Greek word for flushing). In its mild form, the reddening begins four to fourteen hours after exposure and reaches a peak in ten to twenty-four hours. Often we can also see temporary, immediate flushing. Mild sunburn declines soon afterward, and in such cases of what physicians call a minimal erythemal dose or MED, some cells are killed or rendered abnormal. With increased doses, more and more cells are killed and the symptoms become progressively more pronounced until at some point, usually five to ten MED, the skin becomes red and very warm, accompanied by pain and blistering (edema).

All types of skin exhibit erythema upon exposure, but when little pigment is present the effects are worse. For light-skinned people MED may follow as little as ten to twenty minutes of exposure to noon sunlight on a clear day, say, in southwestern United States. Blacks with highly pigmented skin may require ten or more times the exposure to suffer erythema.

Sunburn can also be induced by artificial sources of ultraviolet radiation—for example, by exposure to carbon and mercury arcs. Oxyacetylene torches, used in welding and in glass blowing, induce emission of short-wavelength ultraviolet radiation that produces "sunburn." To different degrees, sunlamps simulate sunburning radiations of sunlight. Enthusiasts who want year-round tan use sunlamps when they cannot get to the beach or during seasons of little sunlight, but excessive exposure to sunlamps can be just as harmful as being out too long in the sun.

Action Spectrum of Erythema

While the skin may flush briefly from sunlight streaming through window glass, the resulting erythema cannot be compared with that caused by direct contact with sunlight. Remember, it is the short-wavelength end of the sun's spectrum, the UV-B radiation, which does not pass through window glass, that causes serious cases of sunburn. But type UV-A rays can also do damage. However, to be effective, UV-A doses need to be much greater, some three hundred to one thousand times greater than UV-B doses in order to induce erythema. Figure 5:1 compares the relative effectiveness of various parts of the ultraviolet spectrum in causing erythema. Wavelength 297 nm is the most effective and at longer and shorter wavelengths the sunburn effect weakens. The detailed analysis of the action spectrum varies, depending on how and

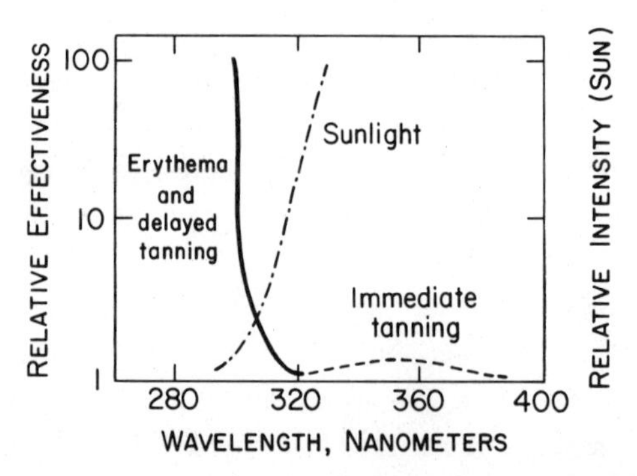

FIGURE 5:1. Action spectra for erythema and tanning. The curves are approximate and vary with the area of skin irradiated, the time at which the observation for erythema is made (here, twenty-four hours postirradiation), and the purity of the light (here, use of a double monochromator). The curves have been interpreted as a result of action on both epidermis (especially for the shorter wavelengths tested) and the dermis (especially for the longer wavelengths). Data for action spectrum from F. Urbach, personal communication).

when measurements are made. Such considerations as the area of skin examined, how long after exposure readings were made, and the purity of the light are all important in making an assessment. Double monochromators are presently almost always used in determining an action spectrum. All studies show, however, that no matter what the minor variations are, it is mainly the UV-B rays that induce erythema at a peak near 300 nm. Some scientists believe, as suggested in Figure 5:3B, that erythema is caused by a composite of effects on the epidermis, mainly by UV-B rays, and on the dermis, chiefly by UV-A rays.

Because fog and cloud cover absorb the infrared or heat rays from sunlight, a pleasant outing at the beach can be overextended. Without the heat to cause discomfort, sunbathers remain at the beach under the illusion that they are not overexposing themselves. The unsuspecting sunbather is unaware that even though smog-free water vapor and water droplets attenuate the intensity by scattering ultraviolet rays, ultraviolet radiation is still reaching the skin. Often, these people are severely sunburned. Even under an umbrella, scattered and reflected UV-B radiation may cause sunburn after excessive exposure. However, smog attenuates UV-B radiation in proportion to its content of radiation-absorbing substances.

How the Skin Gets Sunburned

Vertebrate skin, including human skin, is made up of two layers with each layer composed of cells and cell products. As we see in Figure 5:2, the outermost bloodless layer, the epidermis is made up primarily of cells, while the inner layer, the dermis, contains blood vessels and is made up primarily of fibrous cell products. In the epidermis, five types of cells are distinguishable in the four layers.

Even though the horny epidermal layer composed of dead flattened cells blocks some of the sunlight from passing through, both by absorption and scattering from its flaked granules, considerable ultraviolet radiation near the peak of erythemal power (300 nm) reaches cells of the prickle layer. As shown in Figures 5:3A and B, some even passes through all the layers of the epidermis to the dermis below. When the prickle cells absorb a dose of one or more MED, some are killed and what is commonly known as sunburn is the result. When a large number of cells are

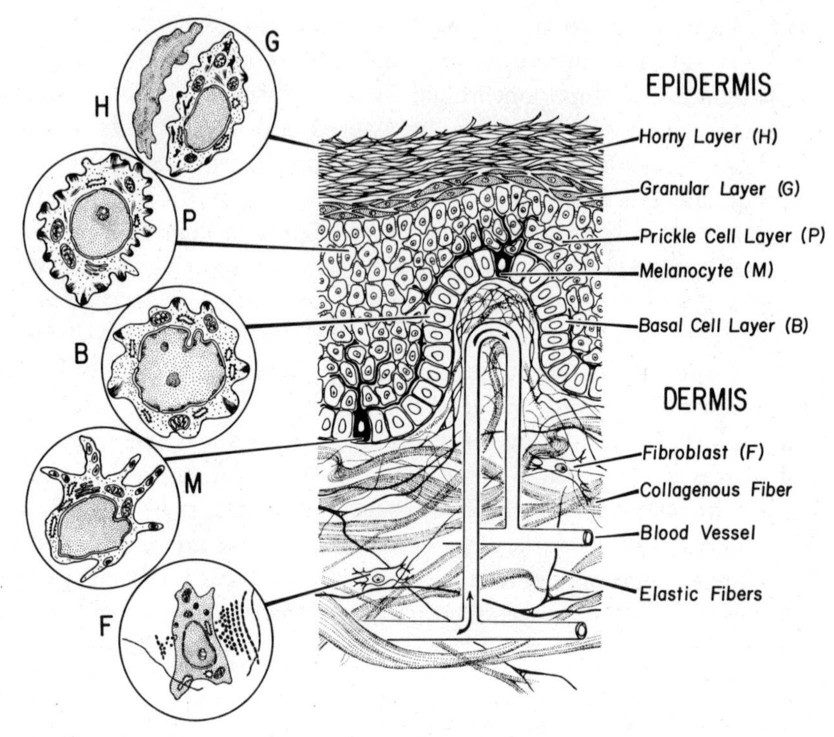

FIGURE 5:2. Structure of human skin as seen in cross section (highly diagrammatic) showing the main types of cells found in the epidermis and dermis: The horny cells (H), continuously being shed from the surface of the epidermis, which are filled with granules of keratin; the granular cells (G), in the process of filling with keratin; the prickle cells (P), named for the interconnections between cells, which are living, active cells; the basal cells (B) which are germinative, giving rise to all the other cells in the epidermis. A basement membrane (not labeled) separates the epidermis from the dermis. In the dermis are shown collagen and elastic fibers secreted by fibroblasts (F). The fibers are formed outside the cells from small subunits secreted by the fibroblasts. The blood supply in the dermis consists of tufts of capillaries, not tubes such as diagrammed. Cellular details largely after Lentz, *Cell Fine Structure,* Saunders, Philadelphia, 1971.

affected after a dose of several MED, tissue fluid and white blood cells infiltrate the epidermis and form a blister. Eventually the fluid is reabsorbed and all the cells above the dead prickle cells are shed in scales, or are peeled off in a layer, familiar to almost everyone who has ever been sunburned. Meanwhile new cells proliferate rapidly from the germinative cell layer as a new thicker epidermis replaces the discarded layer.

The UV-B radiation that damages prickle cells also affects the dermal blood vessels, probably by direct absorption of UV-B and UV-A

A

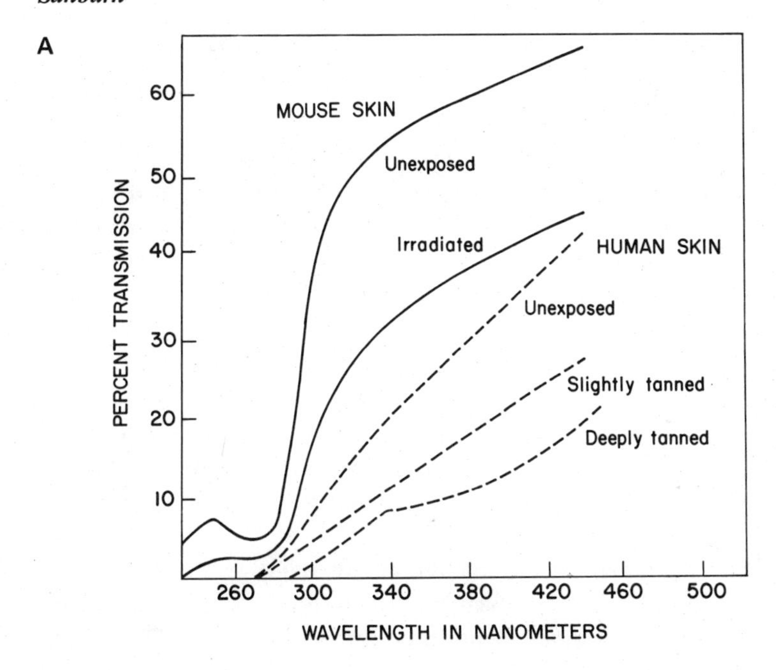

B

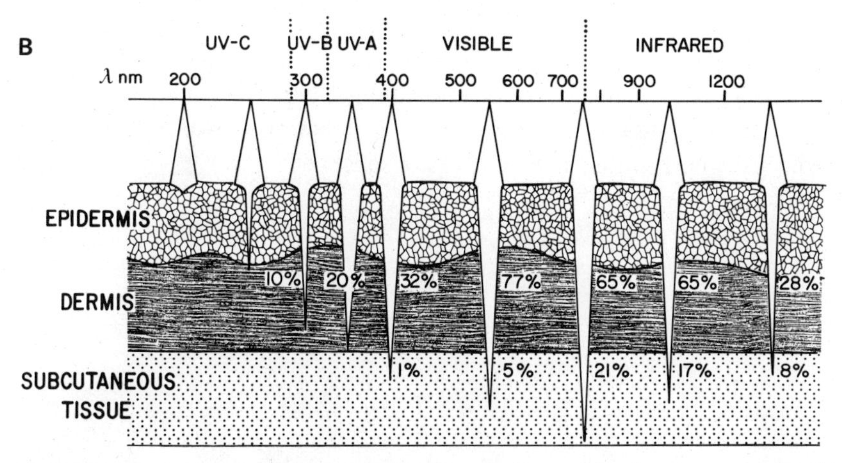

FIGURE 5:3. (A). Transmission of radiation through the ear skin of a young mouse and through the whole epidermis of the human forearm before exposure to ultraviolet light, and reduced transmission after exposure of both to such radiation that induces tanning in human skin. After Kirby-Smith, Blum, and Grady, *Journal of the Cancer Institute* 2 (1952), pp. 403-412. (B). Differential transmission in various layers of the skin. Note that considerable ultraviolet radiation reaches the dermis and that longer wavelengths even penetrate the subcutaneous connective tissue. After Ippen, in *Biological Effects of Ultraviolet Radiation,* edited by F. Urbach, Pergamon Press, Oxford, p. 683.

rays, or by indirect penetration of some diffusable and as yet unidentified substance exuded by injured prickle cells. The walls of arterioles, capillaries, and venules relax and engorge with blood. Erythema and heating of the skin ensues. Simultaneously, sensory receptors in the skin become injured, leading to pain, especially when the skin is touched. The graph in Figure 5:4 indicates that it takes about four days for erythema to recede completely. Sunburn is a pathological condition that leaves cumulative damage in both the epidermal and dermal skin layers.

Thickening of the epidermis, chiefly in the horny layer, gradually wears off, and within six weeks it is probably no thicker than it was before. In very light-skinned people and albinos, who tan little or not at all, epidermal thickening is their main protection after exposure. For a time following a previous exposure, such individuals are somewhat less sensitive to sunlight inasmuch as the thickening protects them. For most people, this effect is of secondary importance.

There is another—as yet not fully explored—system that may protect the skin. We know that urocanic acid, a product of the deamination of the amino acid histidine, found in epidermal cells in the trans form (Figure 5:5) absorbs sunburn ultraviolet radiation. In so doing, it is converted to the cis form. The cis form then reverts back to the trans form, permitting a cyclic dissipation of some of the UV-B radiation. If effective, and increased in quantity with sunlight exposure, it might explain some of the increased resistance to UV-B radiation shown by individuals who have been sunburned and do not tan.

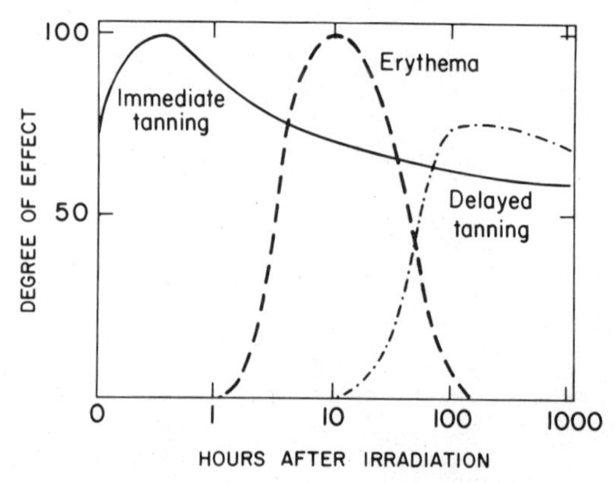

FIGURE 5:4. Time of appearance of erythema and pigmentation. After Hamperl, Henschke, and Schulze, *Virchow's Archives* 304 (1939) pp. 19-33.

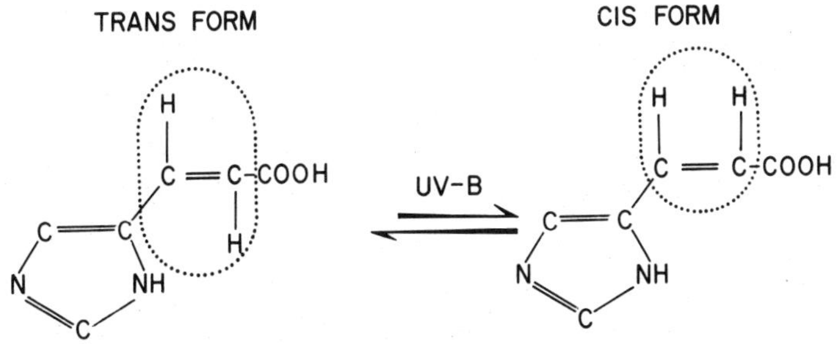

TRANS FORM CIS FORM

UV-B

FIGURE 5:5. Urocanic acid, derived by enzymatic removal of the amino group from the amino acid histidine, exists in two forms, cis and trans. In the stable trans form, the hydrogens are on opposite sides of the double bond. UV-B radiation converts the trans form to the cis form in which both hydrogens are on the same side. The cis form reverts spontaneously to the trans form.

Sunlight affects dermal cells since some ultraviolet radiation passes through the epidermis. But sunburn rarely acutely damages the dermis beyond temporary enlargement of blood vessels. Chronic injury develops only after chronic overexposure.

Tanning

Skin tans are familiar to all light-skinned people. Tanning also occurs in those with dark skins, but it is less evident. There are two types of tanning, immediate and delayed. Immediate tanning results from oxidation of precursors of melanin present in skin cells. Its action spectrum is very broad, covering both UV-A and visible light span in sunlight. Immediate tanning even occurs through ordinary window glass. Since the amount of precursor present in the skin is small, the amount of pigment deposited is likewise limited. Bleached pigment from a previous sunburn also serves as a precursor for immediate tanning.

In Figure 5:4 we see that delayed tanning begins about ten hours after sunburn and reaches a peak in about four to ten days. It then gradually fades because of shedding of pigmented cells, with the whole epidermis being renewed by the basal germinative cells every 30 days. Most tan is gone in a few months, but a residue remains; once skin is sunburned, it is altered irreversibly. The action spectrum for delayed tanning is approximately the same as for erythema. Skin tans result from the activities of cells called melanocytes that, during early development, come from the apex of the embryonic nerve cord and wander to various tissues of the body, where they deposit pigment. About one cell in ten in the germinative layer of the epidermis is a melanocyte. Pigment is deposited actively after a sunburn. Pigmentation may also follow other types of injury to the skin—mechanical and heat, for example.

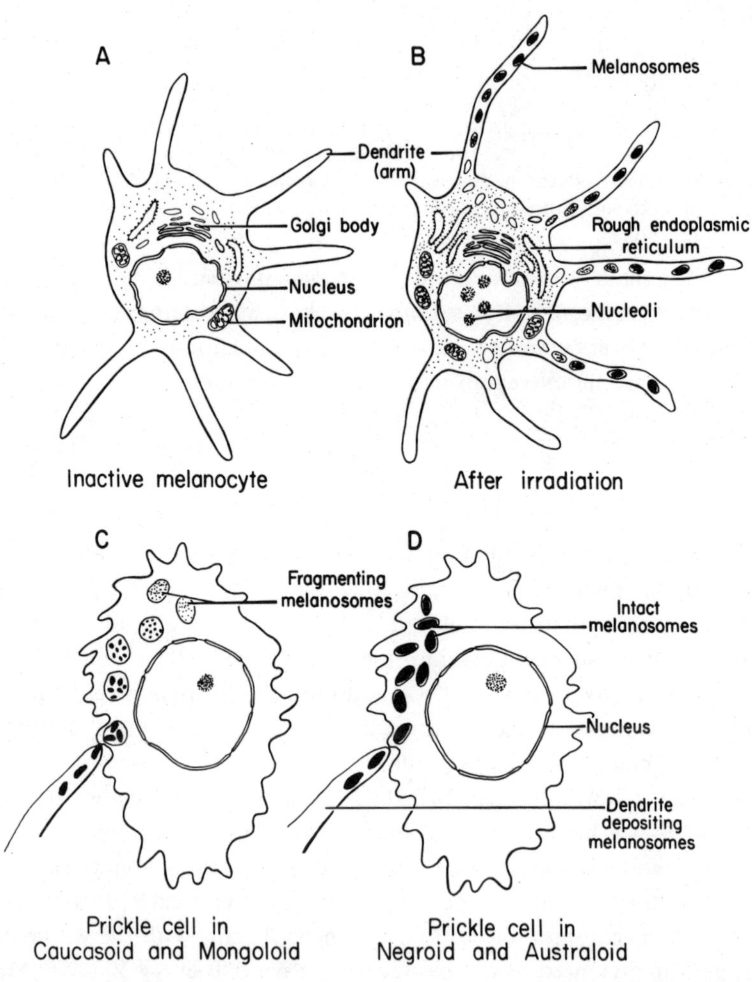

FIGURE 5:6. Diagram of an inactive (A) and UV-activated (B) melanocyte. Note increase in synthetic organelles, rough endoplasmic reticulum and Golgi body, in B and the development of many melanosomes. Pigmentation proceeds as the melanosomes move into the dendrites for transfer to the prickle cells. (C) and (D) are diagrams of prickle cells showing transfer of melanosomes from a dendrite in the two types of skin. See text for details.

The melanin of melanocytes is synthesized in small bodies (cell organelles) called melanosomes, products of the rough endoplasmic reticulum and the Golgi body, which are shown in A and B in Figure 5:6. The melanosome, which has a laminated structure containing the enzyme(s) for the synthesis of melanin from the amino acid tyrosine, gradually becomes filled with melanin. Meanwhile the melanosomes move into the arms (dendrites) of the melanocytes from which they are either secreted into the prickle cell in contact with the dendrite or engulfed by the prickle cell. In light-skinned people, several melanosomes are aggregated into a membrane-enclosed vacuole on entering the prickle cell. Then they are decomposed in the vacuole into smaller fragments that are ultimately released. In dark-skinned people, the melanosomes are not only about twice as large, but they enter singly and do not fragment as in the prickle cell, as shown by C and D in Figure 5:6.

One melanocyte supplies thirty-six prickle cells, with which it forms an epidermal melanin unit. The unit responds to hormones that may increase pigmentation, and has inhibitory feedback control on the hormonal system.

In immediate tanning the melanocytes present are stimulated to develop melanin from melanin precursors already present in the cells, but the melanocyte itself produces no new melanosomes. In delayed tanning initiated by exposure to ultraviolet radiation, the melanocyte becomes synthetically active as shown by increase in nucleoli, endoplasmic reticulum, and Golgi body, and produces new melanosomes. These are colorless at first, but pigment as they move into the dendrites for deposit into prickle cells.

Curiously, despite differences in color among different people, the number of melanocytes in all skin is approximately the same. More melanocytes are active in black than in less pigmented skin and fewest in albinos, such as the "moon children" found among the Cuña Indians of Panama (Figure 8:8). Albinos occur in all races of man and in many species of animals.

The inheritance of pigmentation in mammals, including humans, and in other animals is rather complex and dependent on multiple hereditary factors. For example, in a cross of a black and a white mammal, in addition to offspring intermediate between the parents, they also produce offspring with mosaic coats, a condition in which patches of the coat may be black while others are white. Mosaics may result from the inactivation of genes in some chromosomes. Among humans, this occurs in white patches of skin lacking melanocytes, a condition known as piebald skin, vitiligo, or leucoderma. Vitiligo occurs in skin of all shades of color from black to light-skinned. Treatment with a photosensitizing dye such as psoralen, followed by exposure to UV-A radiation, leads to repopulation of the white areas by division and migration of melanocytes from hair follicles.

Melanin develops by a stepwise oxidation of the amino acid tyrosine in the presence of the enzyme tyrosinase and the subsequent polymerization of the intermediate products. Most of the tyrosinase in the melanocytes appears to be in an inactive form, perhaps coupled to an inhibitor. UV-B activates the enzyme in some way, possibly by removal of the inhibitor. Six successive stepwise oxidations give rise to a series of progressively more fully oxidized compounds. Melanin appears to be a complex mixture of polymers formed from several of these intermediate stages in oxidation. Perhaps the strong absorption by melanin of wavelengths of radiation over the entire ultraviolet and much of the visible spectrum as shown by the dark color (brown to black), is a characteristic of a random polymer with many types of subunits. No one yet quite knows how the reactions operate, and no one has yet identified the exact structure of melanin.

Melanin is ideally suited to its role as a sunscreen in the skin because it not only absorbs ultraviolet rays over the entire spectrum, but also because in light-skinned people it is introduced into the prickle-cell layer of the epidermis in the form of minute particles. Fortunately, it not only absorbs radiation more effectively than if it were in the form of large flakes, but it also scatters light to a greater degree. Furthermore, some of the pigment is arranged as a cap over the nucleus of cells, thereby giving maximum protection to the DNA.

A good tan reduces the ultraviolet rays penetrating to the dermis by about 90 percent. Of course, excessive exposure to sunlight still leaves victims with a sunburn. Even black-skinned people can suffer sunburn from prolonged exposure to the sun inasmuch as their pigment serves only to attenuate the UV-B radiation.

As we all know, tans acquired from being out in the sun disappear after a few months, and for those who cherish it, it seems much shorter. Melanocytes supply melanin granules to prickle cells for only a few days after prolonged exposure, whether or not there is sunburn, becoming less and less active after ten days. Thereafter, in light-skinned people the germinative layer continues to give rise to prickle cells, but they receive little if any pigment. As cells with pigment are pushed upward during the natural growth process, they ultimately become flattened and cornified and are scaled off. Thus, much of the tan, even if acquired by vacationers who work hard at it, will wear off unless they get out in the sun from time to time. We all know that during a stay indoors away from the sun characteristic pallor develops. Every day by just moving in and out we get enough sunlight on our faces, necks, arms, and hands to get some tan.

Delayed tanning also follows exposure to large doses of UV-A radiation, much as does erythema. The mechanism of action may be different for A and B rays but little is known about it.

No one yet knows whether the sunlight that reaches the melanocytes excites their activity directly, or whether an activator diffuses from injured prickle cells to the melanocytes. The delay in tanning after sunburn injury suggests diffusion of a substance from light-damaged cells, but no such substance, even though scientists have tried to find it, has been isolated. It is also not known how the epidermis retains pigment residue in those people who have once acquired it. Strangely, the reflectance and absorbance of such skin is different from that of the skin not previously exposed to the sun.

Variation in Skin Sensitivity to Sunburn

The sensitivity of human skin to sunburn varies widely. Black skin, copiously supplied with melanin, is many times more resistant to UV-B radiation and sunburn than light skin. Resistance decreases as the amount of natural pigment decreases. Even those with light skin exhibit differences in sunburn susceptibility. Those who tan are more resistant than those who freckle or form no pigment at all. Among the latter, those of Scotch, Irish, Welsh, or Breton extraction—in other words, the Celts—are especially susceptible to sunlight. Obviously, albinos are even more so, even though in both groups a little protection is afforded by a thickening of the epidermis. Frighteningly susceptible to the sun's rays are those who suffer from the disease xeroderma pigmentosum. Regardless of initial pigmentation, their skin is quickly burned and badly damaged from even a single overexposure to the sun (see Figure 4:13B).

Intracellular and Molecular Changes in Sunburn

Some evidence suggests that in epidermal prickle cells the minute intracellular organelles, called lysosomes, within a single bounding membrane (see Figure 4:5), rupture during sunburn. Lysosomes retain the various enzymes capable of digesting cell constituents. Presumably, ultraviolet radiation disrupts lysosome membranes, releasing the enzymes within the prickle cells. Release of degradative enzymes from lysosomes activates other enzymes that catalyze synthesis of the keratin granules, not only in the granular layer but also in the prickle cells bordering the granular layer.

Laboratory evidence reveals that enzymes contained in lysosomes disappear from prickle cells, showing that the lysosomes break down. Also when drugs such as the steroid hydrocortisone are administered, keratin granule formation is inhibited in the prickle cells. Generally, this stabilizes lysosome membranes in cells subjected to injurious agents. Without stabilization of the lysosomal membranes, the radiation-injured prickle cells accumulate keratin granules, then die and become flattened cells in the outermost layer of the epidermis.

Not all cell membranes are as easily damaged by sunlight as lysosome membranes. Thus, the plasma single membrane bounding the entire cell and the double membranes of mitochondria are not visibly affected at the time when lysosomal membranes presumably break down. Of course, when cells are given lethal doses of ultraviolet radiation, all membranes are disrupted and abnormal mitochondria are seen in dying cells.

The erythema action spectrum, though suggesting nucleic acid (presumably DNA), does not constitute positive identification of the receptor for ultraviolet radiations in the prickle cells, because it is determined on the basis of light incident on the outer surface of the epidermis rather than on the prickle cells. The intensity of various wavelengths striking the prickle cells is not known because the light entering the prickle cells is first filtered through the horny and granular cell layers that absorb considerable ultraviolet radiation. However, when most of the horny layer is stripped off with adhesive tape before exposure, the action spectrum is not significantly different.

In the bactericidal effect, however, we have seen that the action spectrum clearly indicates nucleic acid as the receptor of the radiation.

Since DNA synthesis, necessary for replication of the chromosomes and cell division, is the reaction most sensitive to ultraviolet radiation, DNA is considered the most likely primary receptor for the radiation (see Chapter 4). Since DNA plays a central role in all cells, the same reasoning might be presumed for the ultraviolet radiation damage of living skin cells.

Modern tracer methods provide direct evidence that DNA synthesis is the process in human and mouse epidermal cells most sensitive to sunburn radiation. Radioactive tracer compounds that are selective for one of the three major macromolecules in the cell (DNA, RNA, proteins) but not for the other two are applied to cells in tissue culture, or injected into the skin and thus supplied to the cells in turn. When the cells are irradiated with doses sufficient to produce sunburn, DNA synthesis is blocked more readily than RNA synthesis or protein synthesis, although it continues in the basal cells of the similarly treated but unirradiated controls. Cells in the basal layer of the irradiated human skin begin to divide again after seven hours and in forty-eight to seventy-two hours there is a marked acceleration of DNA synthesis and cell division. Remember that thickening of the irradiated epidermis in humans begins about forty-eight hours after irradiation and is evident even a week after exposure. This is reflected in the divisions detectable by tracers. RNA synthesis in the basal cells of the epidermis was blocked by larger doses of ultraviolet radiation than DNA synthesis, and protein synthesis only by still larger doses.

Findings from tracer experiments for skin are no different from those obtained for irradiated bacteria, fungi, protozoans, and mammalian cells in tissue culture. Therefore, it is likely that the events that transpire in sunburned skin cells are generally like those in UV-irradiated cells. In other words, the injury occurs in the nucleus or, more precisely, to the DNA. Protection of the nuclear components of skin cells by a melanin cap is, thus, an adaptation admirably fitted to shield from ultraviolet radiation the most sensitive elements in the cells of the skin.

Molecular biologists analyzing sunburn use basal cells extensively because these cells undergo periodic cell division. Cell division, a sensitive indicator of ultraviolet radiation action, is rare in the other cell layers of the epidermis. While the basal cell layer is affected, it is hardly damaged in mild sunburn of human skin because the ultraviolet rays are filtered through horny, granular, and prickle cell layers. Because mouse skin is much thinner than human skin and the basal cells are much more readily reached, mouse, rather than human, epidermis is used in most of these experiments. Obviously, it is also less cumbersome and does not pose ethical problems.

Recovery from Sunburn

Ultraviolet radiation affects epidermal cells by damaging their DNA. The damage probably consists primarily of the formation of thymine dimers, such as are formed during exposure to ultraviolet radiation in all types of cells, including mammalian cells in tissue culture. As we have seen in Chapter 4, thymine dimers, once formed, can be removed in most cells either by photoreactivation or dark repair. However, the photoenzyme needed for photoreactivation has been found only in mammalian white blood cells and fibroblasts, but not in any other mammalian cells tested. On the other hand, there is good evidence for dark repair in mouse and human epidermal cells by the cut and patch mechanism (excision repair) as well as by postreplication repair. It is thought that the damage not subject to repair may be a type of damage in DNA or damage to other macromolecules in the cell, for which no recovery mechanism is known.

Scientists recognized the importance of dark recovery in cells of normal skin when it was discovered how severely damaged the victims of xeroderma pigmentosum became by even one overexposure of the skin to the sun. These people have skin cells with either deficient excision or postreplication repair mechanism (Figure 4:13B). There appear to be five genes involved in regulation of dark repair in human skin cells, but their function has not been elucidated. More conditions than the degree of resistance to radiation are probably regulated by these genes, because in some xeroderma victims organs other than the skin are defective. Ultraviolet radiation resistance in bacteria is also controlled by a number of genes, and a corresponding graded resistance to radiation has been demonstrated (Figure 4:13A).

Some scientists have claimed that the sunburning activity of UV-B radiation is amplified by the simultaneous or subsequent application of UV-A radiation. They suggested, with some evidence, that UV-A radiation reduced the activity of the dark repair system, probably indirectly. However, subsequent experiments have shown only that skin damage from UV-A radiation is essentially additive to that of UV-B radiation. As in the bactericidal effect (Figure 4:3), the dose of UV-A radiation alone required to damage cells is of a different order of magnitude from that of UV-B radiation.

Protecting the Skin from Sunburn

Since some people must work outdoors for prolonged periods, while others enjoy it in their leisure time, lotions and ointments have been developed as screens to protect the skin. A sunburn protective preparation contains chemicals that absorb or scatter UV-B radiation by attenuating the radiation reaching the skin, thus considerably delaying sunburn. The chemicals are dissolved in a solvent or incorporated in an ointment such as vanishing cream, an oil in water emulsion.

While a wide variety of substances absorb the UV-B portion of sunlight, not all of them may be used as sun screens because some irritate the skin. Among the most satisfactory sun screens are the para-amino-

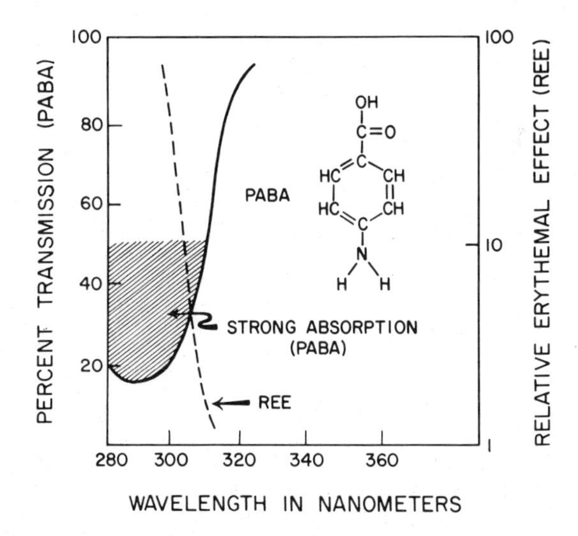

FIGURE 5:7. Effective screening by para-aminobenzoic acid (PABA) against sunburn radiation. Note that absorption is maximal in the very region where the sun's radiation has maximal erythemal effectiveness. Data for para-aminobenzoic acid from Daniels, Van der Leun, and Johnson, *Scientific American* 219 (1966), pp. 39-46. Data for erythemal effectiveness (REE), F. Urbach, personal communication.

benzoates (PABA) and their esters; the ortho-aminobenzoates or anthranilates; and the orthohydroxybenzoates or salicylates, families of related compounds with a variety of substitutions for the hydrogens in the benzene ring (see Figure 5:7). At present, para-aminobenzoic acid (5 percent alcoholic solution) is widely used as a sun screen because it is colorless and blocks out about 95 percent of the UV-B radiation, and seldom provokes undesirable allergic reactions in the skin. It can be applied in solution or in a variety of ointments. It is invisible after application in solution but quite effective, because some of it is absorbed by the skin and is not easily removed even by sweating or washing, except immediately after application.

Though much larger doses of UV-A than UV-B radiation are required to damage cells, the intensity of the former in sunlight is much greater than the latter. This makes it desirable to protect against UV-A as well as UV-B radiation. For this purpose a sun screen absorbing UV-A radiation (e.g., menthyl anthranilate, Figure 5:8, or benzophenone)

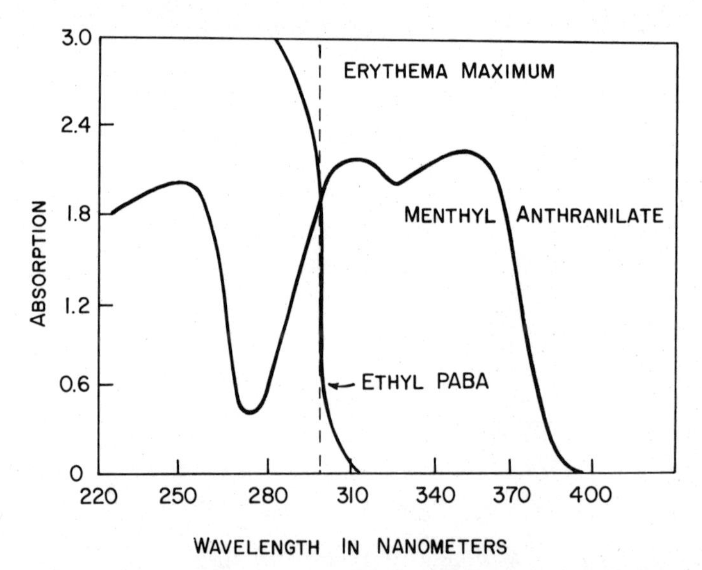

FIGURE 5:8. Absorption by a sunscreen (menthyl anthranilate) for the UV-A part of the spectrum (0.01 percent solution in 95 percent ethyl alcohol). Plotted also is the absorption spectrum of ethyl PABA, an excellent sunscreen for UV-B and UV-C radiation. In this figure the light removed by the solution is plotted, rather than the light transmitted by it, as plotted in Figure 5:7. After Giese, Christensen, and Jeppson, *Journal of the American Pharmaceutical Association, Science Edition* 39 (1950), p. 30.

should be added to the lotion or ointment. Those who work where they are exposed to arc light with UV-C radiation should include screens (e.g., some substituted PABA, Figure 5:8) against this range of wavelengths as well.

In the past many dark pigments were used that absorbed not only UV-B and UV-A radiation but also a goodly part of the visible spectrum —for example, ochre and burnt umber pigments in a cream base. Primitive tropical and subtropical peoples and those at high altitudes and deserts still apply such pigments in grease or even a coat of mud to the skin to protect themselves from sunburn.

Scattering agents such as zinc oxide and titanium dioxide are sometimes incorporated in sunburn preventives, but magnesium oxide, talc, chalk, and kaolin (clay) have also been used. Scattering agents are equally effective for UV-A, UV-B, and UV-C radiations.

It is estimated that the usual maximum exposure to summer sunlight on a clear day would be about twenty MED. Since no one is likely to stay out in direct sunlight all day, a sun screen should protect against at least ten MED although many on the market do not. To avoid a painful burn during a prolonged exposure, the protective preparation should be applied several times because sweating, washing, and abrasion are likely to remove some of it. People of high-risk skin would do best to stay out of the sun at its height between 10 A.M. and 3 P.M.

Effects on the Eye

Human skin does not perceive ultraviolet radiation, but in our eyes it produces an unpleasant sensation of fuzzy bluish light because the cornea and lens fluoresce in the radiation. If a lens transmitting UV-A radiation is substituted for the normal lens lost in an accident or a cataract operation, we can see and even resolve details. The retina of our eyes perceives UV-A radiation but our optical system is unsuitable for vision. The eyes of insects are sensitive to the ultraviolet-A portion of the spectrum and many find nectar by responding to "flower guides" reflecting ultraviolet light, or recognize their mates by ultraviolet reflectance patterns that our eyes do not perceive.

Excessive exposure to sunlight ultraviolet radiation can damage the human eye. It may kill the cells in the outermost lining of the eyeball, and when these cells die, they become opaque and the individual becomes snow-blind. As the dead cells begin to shed, the eyelid is less well lubricated and all eyelid movements become exceedingly painful. The individual also becomes photophobic (hating the light), perhaps because light stimulates slight movements of the eyelids. Recovery from snow blindness is fairly rapid. A new layer of cells appears within a few days, and within a week the eyes return essentially to normal. The problem is particularly serious when snow covers the ground, since reflectance increases the amount of ultraviolet radiation reaching the eye (reflectance from snow is 85 percent; from sand, 17 percent; and from soil or grass, 2.5 percent). In snowfields at high altitudes or in the Arctic, the unshielded cornea may be damaged after being exposed for only a few hours. Eskimos, native to the Arctic, used wooden slit goggles that they have developed to protect against eye damage from the snow.

Polar and alpine animals such as caribous, mountain sheep, mountain goats, ptarmigans, and gulls, in contrast to man, do not appear to be injured by sunlight, even though they are exposed to it directly and reflected from snow. Yet, their eyes appear to be in no way protected from sunlight and are kept wide open to achieve a large visual field. On the other hand, temperate-zone animals, such as the rabbit, opossum, albino rat, mallard duck, and domestic chicken—which, like man, are not adapted to such exposure—develop snow blindness. Absorption of ultraviolet radiation by the corneas of arctic and alpine animals is no different from its absorption by corneas of temperate-zone animals. Either the proteins of the corneas of these animals are more resistant to ultraviolet radiation, or dark repair of damage is much more rapid than in the corneas of temperate-zone animals. The problem remains open.

As arc welders and glass blowers working with quartz have learned, ultraviolet radiation from artificial light sources may also damage the eye. The effect of such exposure resembles snow blindness. The eyes can be readily protected by appropriate ultraviolet-absorbing glasses or hoods.

Cataracts are also believed to result from excessive exposure of the eyes to UV-B radiation, and probably makes a much earlier appearance in individuals exposed to excessive UV-B radiation. When cataracts occur, cloudy or opaque areas appear in the lens of the eye. In the early stages of cataract, the cloudy area consists of denatured protein in the fibers that make up the lens; calcium salts may later be deposited in the fibers. Ultraviolet is well recognized as an effective agent for denaturing proteins and may have such action on the lens, though documentation is sparse. On the other hand, cataract also occurs in aging people with no special history of excessive sunlight exposure.

Strong visible light also damages the eye, inducing degeneration of the light receptor cells (rods and cones), in the retina. When cells in the area directly behind the lens in the region of greatest visual acuity are affected, the individual's capability for resolving detail decreases. Sudden exposure to bright light may also cause ruptures in small blood vessels, which may damage the receptors in the area and form localized blind spots without photoreceptors.

For Additional Reading

Briggaman, R. A., and Wheeler, C. E. "Epidermal-Dermal Interactions in an Adult Human Skin: Role of Dermis in Epidermal Maintenance." *Journal of Investigative Dermatology* 51: (1968): 454–465.

Cripps, D. J., Ramsay, C. A., and Cariner, J. "Effect of Monochromatic UV Radiation on DNA Synthesis with *in Vivo* and *in Vitro* Autoradiography." *Journal of Investigative Dermatology* 58 (1972): 312–318.

*Daniels, F., Jr., van der Leun, J. C., and Johnson, B. E. "Sunburn." *Scientific American* 219, (July, 1968): 39–46.

Epstein, J. H., Bukuyama, K., and Fye K. "Effect of Ultraviolet on the Mitotic Cycle, and DNA, RNA and Protein Synthesis in Mammalian Epidermis *in Vivo.*" *Photochemistry and Photobiology* 12 (1970): 57–66.

Fitzpatrick, T. B., Pathak, M. A., Harber, L. C., Seiji, M., and Kukita, A., eds. *Sunlight and Man,* Tokyo: Tokyo University Press, 1974, pp. 100–110, 165–194, 195–215, 751–765.

Giese, A. C., Christensen, E., and Jeppson, J. "Absorption Spectra of Some Sunscreens for Skin Preparations." *Journal of the American Pharmaceutical Association* 39 (1950): 30–36.

Groves, C. A. "Selection and Evaluation of Ultraviolet Absorbers." *Australian Journal of Dermatology,* 14 (1973): 21–34.

*Johnson, B. E., Daniels, F., Jr., and Magnus, I. A. "Response of the Human Skin To Ultraviolet Light." In *Photophysiology* vol. 3, edited by A. C. Giese. New York: Academic Press, 1968, pp. 139–202.

Johnson, J. A. Czaplinski, P. R., and Fusaro, R. M. "Protection against Long Ultraviolet Light with Dihydroxyacetone-Naphthoquinone." *Dermatology* 147 (1963): 104–108.

Keeler, C. "Cuña Moon Child Albinism," 1950-1970. *Journal of Heredity* 61 (1970): 273–278.

Kettlewell, B. *The Evolution of Melanism; the Study of a Recurring Necessity.* Oxford: Clarendon Press, 1973.

Nachtwey, D. S., ed. "Impacts of Climatic Change on the Biosphere. CIAP Monograph 5. Part 1. Ultraviolet Effects." Department of Transportation. Climatic Impact Assessment Program. Washington, D.C., 1975.

Urbach, F. ed. *The Biological Effects of Ultraviolet Radiation with Emphasis on the Skin.* Oxford: Pergamon Press, 1969, pp. 159–163, 237–249, 251–254, 305–314, 469–471, 681–687, 689–692.

Witkop, C. J., Hill, C. W., Desnick, S., Thies, J. K., Thorn, H. L., Jenkins, M., and White, J. G. "Ophthalmologic, Biochemical, Platelet and Ultrastructural Defects in Various Types of Oculocutaneous Albinism." *Journal of Investigative Dermatology* 60 (1973): 443–456.

6

How Sunlight Ages Skin

That sunlight, more than the passage of time, removes the visage of youth has been known by dermatologists for years, but the rest of us remain oblivious to this dictum. In addition to immediate, acute deterioration and rapid healing of the epidermis following sunburn, more permanent changes are induced in both epidermis and dermis by repeated, excessive, long-term exposure of the skin to the sun (Table 6:1). Often the deleterious effects are not recognized as a consequence of overexposure to sunlight in youth, but more popularly interpreted as a result of aging. Frequently, weather-beaten, wrinkly, furrowed, splotchy, excrescence-marred faces characteristic of extreme old age develop in early middle age or sooner as a consequence of too much sun.

Skin ages when its component parts, the cells and the fibrous connective tissue they produce, age. In our multicellular bodies some cells, such as those of the epidermal basal layer, continue to divide throughout life; others—for example, nerve and muscle cells—differentiate and cease dividing at birth. Nevertheless, they continue to function for a lifetime, with gradually lessening activity and progressive filling with insoluble wastes and pigments. Possibly the waning enzymatic activity results from the cumulative damage to DNA, on which all cell syntheses depend.

One might expect that tissues made up of cells that are continuously replaced would not age. Consider human epidermis where the basal cells divide rhythmically, and when they mature, replace cells in the other epidermal layers monthly. Similarly, fiber-producing cells (fibroblasts) in the dermis divide slowly and continuously replace some of the connective tissue fibers. Yet, the properties of both epidermis and dermis change as we age.

Table 6:1. Effects of Various Doses of UV-B Radiations on Human Skin

Minimal erythemal dose (MED)	5 to 10 MED	Chronic overexposure to sunlight
	Signs and symptoms	
Temporary reddening of skin (erythema) Slight tanning, if any	Deep red coloration Blistering Marked tanning following reddening Pain	Permanently pinkish skin with yellow underlay locally or sometimes widespread Splotchy tanning – each tanned area surrounded by unpigmented areas Flabby skin with dry, leathery surface Deep wrinkles Thinning of skin and change in pattern
	Changes in skin cells and fibers	
Injury to prickle cells Temporary enlargement of dermal blood vessels Possibly slight increase in cell division of germinative cells	Many prickle cells killed Layer infiltrated with tissue fluid (blistering) Peeling of dead layer of cells (desquamation) Stimulation of melanogenesis in melanocytes and "tanning" of prickle cells Stimulation of cell division in germinative layer, causing thickening of epidermis	Killing some cells and damaging of organelles, especially in epidermal cells Damage to blood vessel cells, resulting in flabby blood vessels. Deterioration in melanocyte function; many killed Solar elastosis in dermal connective tissue fibers–altered chemical nature and physical properties of elastic and collagenous fibers Damaged sweat gland cells and sebaceous gland cells

When skin cells age, it is likely that they accumulate damage from internal bodily changes, from the external environment, or perhaps both. Nondividing cells may send "aging" factors into the blood, which carries them to epidermal basal cells and to dermal fibroblasts and other cells. What is more, skin is constantly exposed to the external environment, especially sunlight. Naturally, if genetic messages in basal cell DNA are altered in this manner, cells in the other epidermal layers derived from the basal cells would also change. Fibroblasts, similarly injured, could produce abnormal fibers. While biologists are still puzzled by aging factors, enough is known about the action of sunlight on our skin to explore the results of chronic overexposure.

Pattern Changes

In some people, chronic overexposure to the sun results in thinning of skin, as seen in a cross-section presented in Figure 6:1. (Remember that the thickening of the epidermis after a sunburn is temporary.) It comes in good part from a change in the overall pattern of the epidermis and dermis. Some areas of normal skin show downward fingerlike extensions of the epidermis (epidermal ridges) that alternate with upthrust fingerlike extensions of the dermis (dermal papillae). In section, the line of contact between epidermis and dermis exhibits a wavy pattern, as we have already seen in Figure 5:2. In chronically sunburned skin, the contact line between epidermis and dermis flattens and epidermal ridges and dermal papillae nearly disappear. Blood vessel patterns are also altered. Little pattern change occurs after sun damage in skin areas that normally show no wave pattern. These pattern changes appear to be quite specific. For example, dermatologists have investigated the facial and neck skin of a twenty-five-year-old chronically overexposed to the sun, and found the typical pattern of sun-damaged skin. In comparison, they studied the same areas of skin of a seventy-year-old who was not overexposed to the sun and they discovered a normal pattern, still showing well-developed epidermal ridges and dermal papillae. Not surprisingly, a given person may exhibit altered patterns in sun-damaged areas while protected skin remains normal.

Not every case of sun damage alters the pattern. Sometimes, dermal papillae and epidermal ridges remain even though they may be somewhat modified. Nor does sun-damaged epidermis always become thinner; The number of layers of epidermal cells is sometimes the same as in the normal skin of the same individual. Curiously, the thinning of sun-damaged epidermis may occur in certain skin areas only, while other areas show little or no change. Even when the number of epidermal cell layers does not decrease by sun damage, the layer arrangement may be distorted, and the cells dividing from the basal cell layer may mature more slowly, less synchronously and less completely than normal epidermal

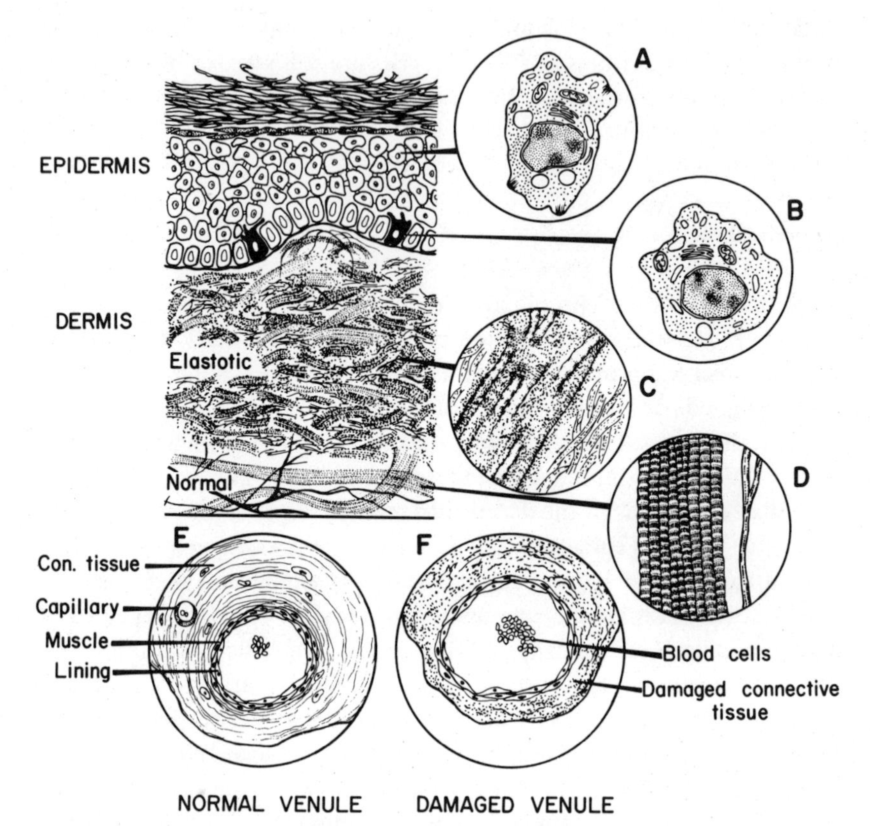

FIGURE 6:1. Diagrammatic cross section of the skin of a person with solar elastosis from overexposure to sunlight. Note the change in boundary between the epidermis and the dermis in comparison to the normal boundary shown in Figure 5:2. Changes in cells and connective tissue with sun damage are shown in inserts (A) to (D) as seen under an electron microscope; and changes in a blood vessel (venule) are shown in inserts (E) and (F) as seen under a light microscope. (A) Prickle cell with altered nucleus and damaged organelles. (B) Melanocyte that has lost its rootlike extensions (dendrites) and its capacity to form melanin in its melanosomes. (C) Elastotic connective tissue fibers in sun-damaged zone. (D) Normal connective tissue fibers, collagen (left) and elastic (right), below the reach of damaging rays. (E) Cross section of a normal small vein (venule). (F) Cross section of an elastotic venule with degenerate connective tissue.

cells. Furthermore, abnormal nuclei appear in some cells probably as a result of radiation damage to the cellular DNA, which then fails to program normal maturation of the cells (Figure 6:1).

Dermatologists call sun-damaged skin elastotic when the elastic fibers multiply and fragment and the collagen fibers observed under the light microscope stain differently from normal. Elastotic conditions in the dermis may appear a decade or more before similar patterns show up in the epidermis. Records often indicate little if any epidermal damage after the first appearance of elastosis in the dermis, or they may show progressive changes in the structure of interlocking epidermal ridges and dermal papillae and thinning of the epidermis only after continued chronic overexposure to the sun. But such differences may be more apparent than real, because it is difficult to record the cumulative sunburn dose administered to the skin in view of the many years of exposure to sunlight involved.

The different ways in which the epidermis responds to sunlight may reflect the way in which different parts of the body are exposed to the skin. For example, buttock skin, which is little if at all exposed to the sun, is likely to remain normal in almost anyone. On the other hand, it is possible that inherent differences in skin sensitivity do exist in different portions of the body.

But the question still remains as to why the epidermis is slower to show sun damage than the dermis. Since the turnover of cells in the epidermis is very rapid, the entire maturation cycle from basal germinative cell to horny cell lasting only twenty-eight days, damaged cells could be shed quickly and replaced by other cells showing little or no damage. By comparison, in the dermis the rate of cell division is very slow. Consequently, the cells could sooner accumulate sun damage that shows up in their synthesis of abnormal (elastotic) connective tissue fibers.

The change in the epidermis of sun-damaged epidermal cells may result from direct action of ultraviolet radiation on the cells, as well as from failure of the damaged dermis to supply normal requirements to the epidermis. Remember that the blood supply for the epidermis is the rich capillary bed in the dermal papillae, as shown earlier in Figure 5:2. When these vessels are damaged, the epidermis lacks adequate supplies.

Melanocytes

When melanocytes are injured, the result is mottling or patchy tanning. Following periodic, excessive sunburn, melanocytes disappear from small areas of skin. Since melanocytes occasionally divide, they may repopulate a damaged area, but after chronic overexposure to the sun, large areas may completely lack active melanocytes. As we see in Figure 6:2, pigmented patches of such skin are sometimes larger and

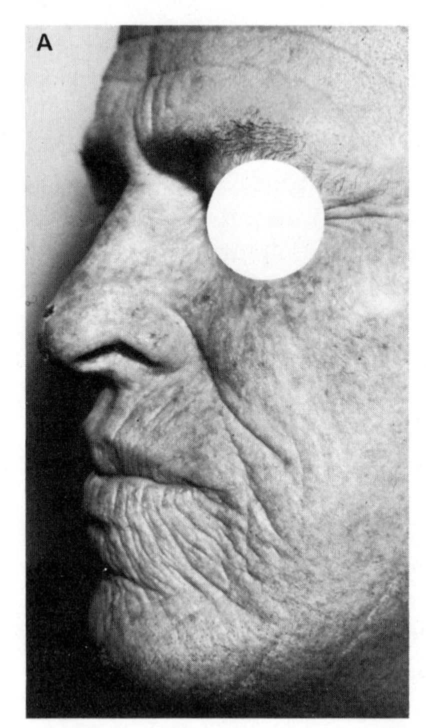

FIGURE 6:2. The skin of individuals with chronic sunburn damage. (A) Note deep wrinkling, splotchy pigmentation, and the warty spot (keratosis) on the nose. Courtesy of Dr. Alvin Cox, Dermatology, Stanford Medical School, Stanford, California, (B) Note deep furrows on the face of the sun-damaged cheek of the individual. Courtesy of Mr. Carl May.

deeper in color than freckles, thus contrasting with surrounding unpigmented areas of the skin.

Under the electron microscope, melanosomes may still be seen in melanocytes in the epidermis of chronically sun-damaged skin, but they have lost the capacity to produce pigment. Biologists believe that the enzyme or enzymes required for melanin synthesis are destroyed and not replaced because of nuclear damage to the cells and the nucleic acids. However, some precursors of melanin may still be present. Since the synthesis of enzymes occurs under the influence of a template supplied by messenger RNA and since the messenger RNA is transcribed from DNA, if the DNA template has been damaged by radiation, no more RNA messages will be sent to the ribosomes to activate the enzymic protein on which melanin synthesis depends. Given the limited lifetime of messenger RNA, when its supply in melanocytes is degraded, no more is available for enzyme synthesis. In turn, the enzyme or enzymes also gradually degrade, terminating the capacity of the melanosome to produce melanin. In the end, only the cell body or central part of the melanocyte containing the nucleus remains, even though it may be somewhat abnormal. Figure 6:1B shows that the dendritic extensions by which pigment is transferred atrophy and disappear. Still another possible change in melanocytes is the formation of giant cells, sometimes bipolar and with two nuclei, which for a time deposit a large amount of pigment.

The population of melanocytes in the skin is fairly large, about 1150 to 2000 per square millimeter in forehead skin and 560 to 1000 in thigh skin. The number of melanocytes per unit area is greater on the face, neck, and forearms than on the remainder of the body, and is about the same in skin of all colors.

Counts indicate that the number of melanocytes in the skin per unit area decreases by about 11 percent every ten years. In older people, the melanocytes also change in size and staining properties. Although dermatologists take care in making estimates of how melanocytes change with age, they do not eliminate environmental factors even when sun exposure has been mild. Sunburn damage may account for some of the decrease in number of melanocytes as people age.

Lubrication

Generally, skin becomes dry and less well lubricated after chronic sun exposure. The skin is lubricated by sebum, a fatty substance secreted by the sebaceous glands, usually attached to hair follicles. As seen in Figure 6:3, the sebaceous gland has an inner lining derived from the epidermis similar to the hair follicle into which it pours its contents. In some areas, such as the lips and mucous membranes, these glands empty directly on the epidermis. The sebaceous glands of chronically sun-damaged skin secrete less sebum. However, cells lining hair follicles and the sebaceous gland seem to escape structural alteration by sunlight. Overlying tissues probably shield them from the damaging effects of sunlight ultraviolet radiation.

Since excessive sunlight damages dermal blood vessels, the blood supply to the sebaceous glands and the sweat glands in such skin is reduced. Dermatologists have not studied this phenomenon enough, but it is possible that the altered pattern of contact between epidermis and der-

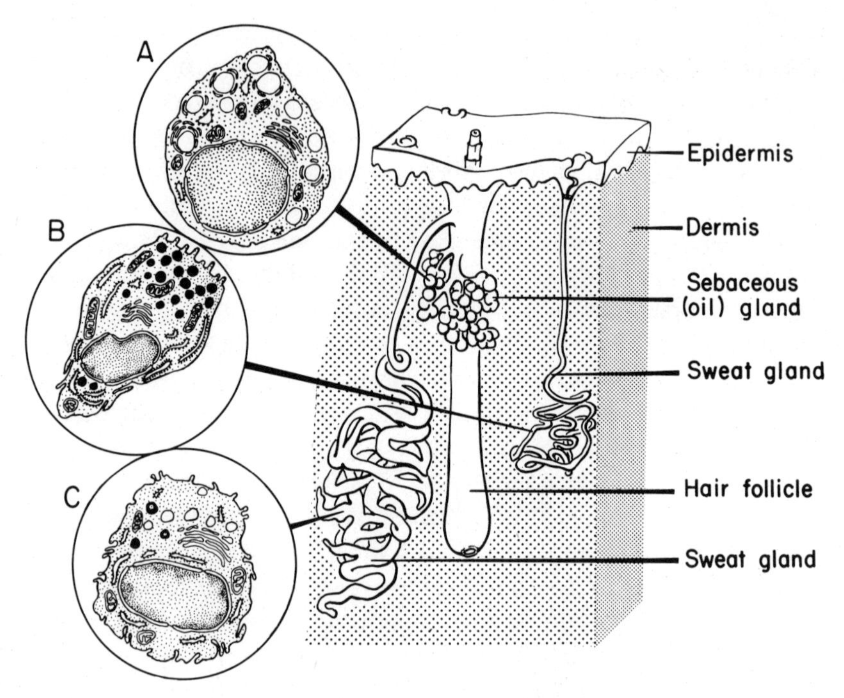

FIGURE 6:3. Diagram of a sweat gland and a sebaceous (oil-secreting) gland opening onto a hair follicle, and another type of sweat gland opening directly onto the surface of the skin. In circles are shown the types of cells that line the glands and the nature of their secretory droplets as they would appear in electron micrographs. Note that in all cases the typical cell organelles are present. (A) Sebaceous gland cell showing large, clear oil droplets. (B) Cell in sweat gland opening onto skin. (C) Cell in sweat gland opening onto hair follicle. Stereogram of gland after Montagna, *The Structure and Function of the Skin,* Academic Press, New York, 1962, p. 376. Cellular details after Lentz, *Cell Fine Structure,* Saunders, Philadelphia, 1971.

mis and the somewhat less well-organized structure of chronically sun-damaged skin impedes lubrication. In sun-damaged skin the disorderly maturation of cells, with abnormal nuclei of variable size in the prickle cells and granular cell layers, to form horny cells may also add to the disorganization, making passage of lubricant through cell layers difficult.

Wrinkling

When skin wrinkles, the surface creases as facial muscles contract or the jaw moves. Excessive exposure to sunlight plays a critical role in premature wrinkling. But sunburn damage to the skin does a lot more than advance the onset of wrinkling, it produces coarser, more extensive, and deeper furrowing than the wrinkling of old age. The dry and scaly surface of repeatedly sunburned skin, which may also take on a leathery appearance, sometimes leads to pachyderm or elephant skin, as shown in Figure 6:4.

In local areas it is possible to abrade the epidermis, and when it is removed, it may be followed by regeneration of more normal-looking epidermis. But it is impossible to reverse the deep-seated effects of chronic overexposure to the sun. These effects pass into the pattern of the epidermis and the dermis, resulting in modification of the structures in the dermis as well. Maintenance of normal epidermis has been shown to depend on the presence of a normal dermis below it.

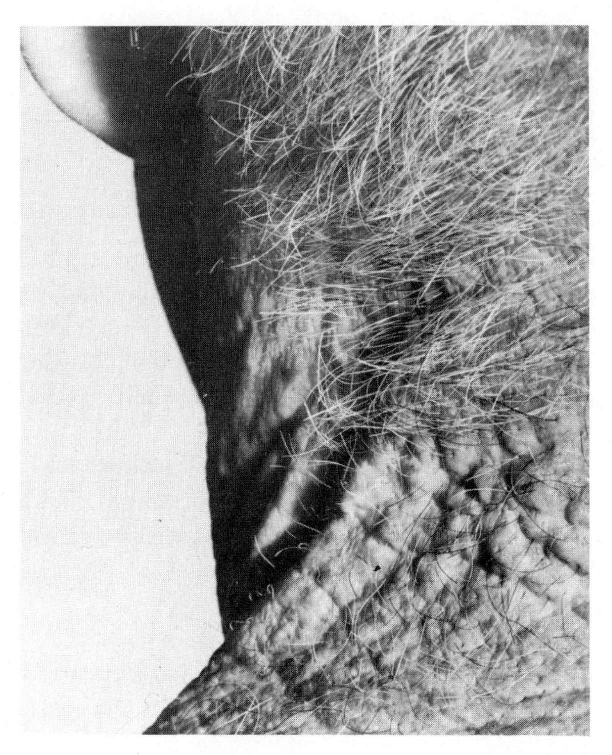

FIGURE 6:4. Pachyderm or "elephant skin" on the neck of a person overexposed to the sun. The age at which this appears depends on the exposure to the sun. Courtesy of Dr. Frederick Urbach, Dermatology, Temple Medical School, Philadelphia, Pennsylvania.

Blood Vessels

The florid skin, pink or reddish colored, seen in the weatherbeaten, sun-damaged skin of sailors, farmers, ranchers, and other outdoor workers, partly results from thinned epidermis more transparent to blood in dermal vessels and partly comes from enlargement of blood vessels themselves. Consider Figure 6:1F. As we see in Figure 6:1E, the blood vessel is a tube with a cavity lined by a layer of elongated cells, surrounded by layers of muscle fibers and connective tissue fibers. Contraction and relaxation of the muscle fibers permits variation in the size of the lumen, while the elastic connective tissue fibers provide elasticity. The muscular wall is more prominent in arterial than in venous vessels. A capillary consists of little more than the lining cells.

With excessive sun exposure, skin blood vessel walls become flabby as their connective tissue deteriorates. In Figure 6:1F, we see the vessels capable of accommodating more blood than normal skin blood vessels. Venous vessels are altered more than arterial vessels.

Connective Tissue

Skin dermis, several times thicker than the epidermis, is made up largely of connective tissue fibers secreted by fibroblast cells that are rather inconspicuous among these cell products. Epidermal cells line the sebaceous glands, the sweat glands, and the hair follicles, all of which develop from the basal cell layer of the epidermis and dip deeply into the dermis. Figure 6:3 shows this arrangement.

The dermis is made up of two intergrading connective tissue layers, one with the fingerlike papillae projecting into the epidermis and another with more linear fibers below it. The lower layer is loosely connected to the subcutaneous tissue below it. The entire skin can be lifted and felt by the fingers as a distinct protective cover.

Connective tissue fibers, collagenous and elastic, are secreted by fibroblast cells. Both types of fiber are proteinaceous, that is, made up of amino acids; both have a large proportion of the amino acids glycine and proline, but elastic fibers have more valine molecules while collagen fibers have more alanine molecules.

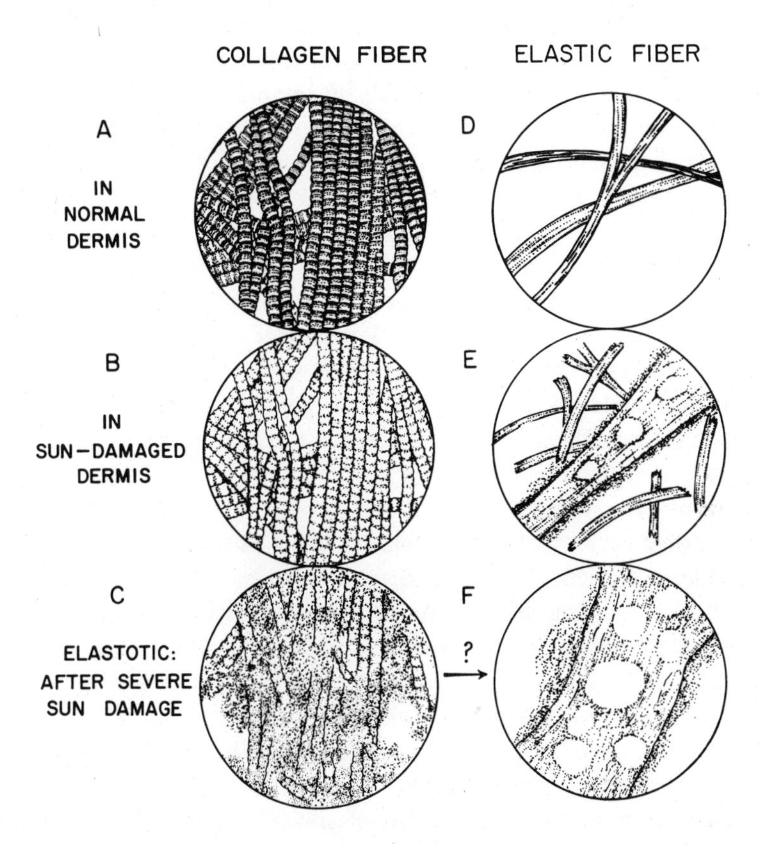

FIGURE 6:5. Diagram of deterioration of the connective tissue fibers in the skin after chronic exposure to sunlight, as seen under the electron microscope. (A) Normal collagen in the dermis showing characteristic periodic markings on the fibrils composing a fiber. (B) Collagen in sun-damaged dermis with less perfect markings on the fibrils. (C) Collagen in skin considerably damaged by sunlight (elastotic) showing tangled fibrils with almost no periodic marking and much formless granular material. Such collagen stains more like elastin in elastic fibers rather than collagen. (D) A normal elastic fiber with nonperiodic lengthwise filaments. (E) Shortened elastic fibers in sun-damaged dermis along with an enlarged fiber staining like elastin, but of unknown origin. One suggestion is that the elastic fibers enlarge by incorporation of other material such as ground substance and degenerating collagen fibers forming the structure seen in (F). Data from Mitchell, *Journal of Investigative Dermatology* 48 (1967, pp. 203-220). (F) Probably a late stage in degeneration of a collagen fiber with no evidence of fibrils and much material staining like elastin of elastic fibers rather than like collagen. Granular material is shown attached to the fiber and large gells inside the fiber.

As drawn in Figures 5:2 and 6:5A and D, the two types of connective tissue fibers appear quite different. Under the light microscope, collagen fibers look like wavy threads on ribbons varying in length from one to twelve micrometers. Figure 6:5A shows them running in all directions. Most of their terminations are indistinct, but in the lower layer of the dermis they tend to be linear. Each collagen fiber is made up of parallel fibrils cemented together, each about half a micrometer in thickness. The fibrils do not branch but the fibers do. Examined under an electron microscope, collagen fibers show bands with repeated units at an average distance of 64 nm from each other. When boiled in water, collagen fibers dissolve into gelatin.

While collagen fibers are flexible, they resist pulling. For human tendon collagen fibers, the breaking point is several hundred pounds per square centimeter. Even when forced, collagen fibers stretch only a few percent of their length and so account for the skin's resilience. For this reason, they were often called inelastic fibers in the past.

Figure 6:5D depicts elastic dermal fibers as thin, single-cylindrical threads or flattened ribbons much smaller in diameter than collagen fibers. They branch and coalesce freely, forming a loose network. Capable of stretching to one and a half times their original length, elastic fibers return to practically their former length when released; their breakpoint force is about one-tenth that for collagen fibers. When massed together, elastic fibers appear yellow.

In the dermal papillary layer, elastic fibers are fewer in number than in the lower layer, but they form a continuous fine layer under the papillae. The collagen fibers pass in all directions and a jellylike ground substance fills the space between the fibers.

In the lower layer of the dermis, elastic fibers form thick networks between the collagen fibers and condense around hair follicles, sebaceous glands, and sweat glands. Collagen fibers are more conspicuous than elastic fibers, and for the most part run parallel to the skin surface, although some run in all directions.

Inasmuch as the dermis consists largely of connective tissue fibers, any change resulting from overexposure to sunlight is evident from changes in the overall fibrous structure of the dermal layer and in the individual fibers themselves (see Figures 6:1 and 6:5).

Considerable sunburn damage occurs from overexposure in the dermal elastic fibers. As they become elastotic they thicken, clump, fragment, and chemically alter. The changes may be seen by staining; they are stained by basic dyes instead of acid dyes as in normal skin. Also, elastic fibers initially increase in number. In places where they mass, they give the skin a yellowish tint either locally or over wide areas and can be seen in some who have been long overexposed to the sun. Elastotic skin shows altered elastic fiber patterns.

Dermal collagen fibers also change during chronic sunburn. In the upper half, they are replaced to some extent by a homogeneous material and unlike normal fibers selectively take up the dye orcein (an alcohol-soluble dye extracted from certain lichens), which indicates chemical changes. In undamaged skin, the dye is usually positive for elastic fibers rather than collagen fibers. As shown in Figures 6:5A to C, under an electron microscope they reveal a loss of the periodic cross-markings characteristic of collagen. Chemical analysis indicates that chemical changes have occurred. A less common form of deterioration seen in such skin is the development from collagen fibers of homogeneous droplets of material, which react with acidic stains. Generally, it appears that in sun-damaged skin some fibrous material is converted to, or replaced by, jellylike material. Undefined chemical changes also occur in the ground substance in which the connective tissue fibers lie.

The degree of elastosis in the skin is closely correlated with its degree of sunlight exposure. A dermatologist studying this problem has suggested turning the tables around—that is, clinically estimating a patient's degree of sunlight exposure by sampling skin elastosis.

When dermal elastic and collagen fibers change because of sun damage, the physical properties of skin likewise change. For example, in normal, unexposed skin, collagen fibers resist deformation and spring back to their original condition as soon as released. They are chiefly responsible for the skin's suppleness. On the other hand, deformed sun-damaged skin very slowly returns to its original state. With the loss of suppleness and resulting flaccidity, wrinkles pass more deeply into the dermis and become more permanent. Sunburn damage to the dermis gives skin its weatherbeaten and leathery look.

Intriguingly, when an investigator grafts a piece of sun-damaged skin to an unexposed, and presumably normal, area of young skin, dermal elements regenerate considerably, suggesting that some changes in sun-damaged skin can be reversed. Unfortunately, no clinical techniques have been found to use this discovery to rejuvenate sun-damaged skin.

In normal skin, dermal fibroblasts slowly and periodically replace both elastic and collagen fibers. How fibers are removed is still not known, even though enzymes specific for hydrolysis of elastin in elastic fibers and collagen occur in the skin, as do ameboid cells that engulf and digest partially decomposed fibers. Remember, as we have seen in Figure 5:2F, connective tissue fibers are self-assembled outside the cell from small subunits manufactured by fibroblasts and secreted extracellularly. These subunits are produced in accordance with instructions given by DNA in the fibroblasts. In sun-damaged skin, the altered dermal fibers probably result from defective subunit synthesis in the radiation-damaged cell. Biologists believe that the cumulative damage to DNA templates of the fibroblast chromosomes, irradiated periodically for years, leads to changed messages. This results in synthesis of altered enzymes that in turn govern the synthesis of fiber subunits. Obviously, if the subunits manufactured by altered enzymes are not precursors to normal fibers, abnormal fibers will be formed.

Normal elastic fibers, drawn in Figure 6:5D, are small in diameter and, when seen under the electron microscope, they show fine linear structures scattered lengthwise. Figure 6:5E reveals what probably happens after progressive sun damage. The linear structures become less evident or absent and the elastic fibers appear to be shorter and more numerous. After very extensive sun damage, elastic fibers, as depicted in Figure 6:5F, can no longer be identified in the dermis, even though the larger fibers present the stain much like elastic fibers.

Compared to elastic fibers, the normal collagen fiber is large in diameter and consists of a group of parallel fibrils. Under the electron microscope, we see in Figure 6:5A regular crosswise markings in register across the fibrils of collagen fibers. Following progressive sun damage, the periodic fibril cross-markings become less evident, as shown in Figure 6:5B; and when they are no longer identifiable, the fibrils, as Figure 6:5C shows, become disordered and tangled. Finally, in extreme sun-damaged cases, dermal fibers show no fibrils and appear to have little or no organization (see Figure 6:5F). These badly damaged fibers stain more like elastic fibers than collagen.

While some investigators believe that these changes occur when collagen fibers incorporate the elastic ones, it is more likely that the information from the damaged fibroblast DNA has become scrambled as the cells are progressively damaged by ultraviolet radiation. This results in the synthesis of subunits for fibers with properties of both elastin and collagen, but without the organization into fibers of either. Chemical determinations of such modified fibers indicate some resemblance to elastin. Because experiments over the prolonged period of time required to produce elastosis would be difficult to perform, and because such evidence as we have is assembled from sections of fixed and stained material from several individuals, the detailed sequence of events in development of elastosis remains largely hypothetical.

One of the mystifying elements in this story is why UV-radiation damage seemingly occurs unevenly throughout the dermis. It is difficult to explain why the fine collagen fibrils in the dermal papillae often stain like normal collagen even though those just below them are elastotic. Either the fibroblasts in this area are protected in some way, are more resistant than those below them, have better DNA repair mechanisms, or have migrated from regions below sunburn damage. And, even in the elastotic portion of the dermis, there appear occasional islands with normal elastic and collagen fibers. Such islands could be products of fibroblasts that have escaped radiation damage and continue to produce normal fibers when their neighbors are producing abnormal ones, or could be migrants from normal dermis. The deepest part of the dermis below the elastotic zone is often quite normal in staining properties and appearance, most likely because the damaging UV-radiations do not penetrate that far (see Figure 5:3B).

Surprisingly, when an elastotic skin is cut or damaged, normal collagen and elastic fibers are laid down during healing. While one could see this as a reversal of the ultraviolet radiation-induced damage or removal of inhibitors, other explanations are possible. Either islands of normal fibers occur among the elastotic fibers, presumably produced by fibroblasts that escaped injury from UV-B radiation or repaired the injury, or normal fibroblasts may have invaded the skin from below the damaged area. These could multiply and invade the wound, producing normal fibers for healing.

Comedones (Blackheads)

Chronic sunburn damage to the skin leads to the enlargement of comedones, commonly called blackheads. These are atrophied sebaceous glands with large cavities. The cavity contains, in addition to a dried plug of sebum and some horny material, a variety of bacteria including one characteristic of acne, as well as some yeasts and sometimes a follicle mite. Some of these organisms are normally present in small numbers on our skin surface, but it is only when a receptacle such as an atrophied sebaceous gland is provided that these organisms invade it and multiply. The body of the gland enlarges to about double the diameter of a pinhead in chronically sunburned skin, because it is not limited by resilient connective tissue as it would be in the dermis of normal skin. Comedones are usually seen on the skin of individuals with thinned epidermis, vacuolated prickle cells, elastotic dermis, and altered skin structure, all characteristic of chronic sunburn damage. While comedones are unattractive, they are no more serious than acne.

Ultraviolet Action Spectra

No action spectrum has been published for chronic sunburn damage to the skin. An action spectrum for such a long-delayed effect would be difficult to determine, and an action spectrum for dermal damage would be even more difficult. The ultraviolet radiation incident on the surface of the skin is filtered by passage through the epidermis, and only about ten percent of the radiation at wavelength 300 nm gets to the dermis, about 20 percent at 350 nm; about 32 percent of violet light at 400 nm reaches the dermis (see Figure 5:3B). It is possible that the UV-A radiation, which only adds to the action of UV-B radiation on the epidermis, may of itself be a major factor in sunburn damage to dermal structures. However, it may only add to the damage of the little UV-B radiation that reaches the dermis. The problem remains open.

Racial and Genetic Differences

Deeply pigmented (black) skin retains a youthful appearance longer than poorly pigmented skin because sun damage is greatly delayed by the screening action of the pigment. With continuous exposure to the sun, black skin ultimately gets "weather-beaten" also. In Figure 6:6 the chart shows how, with the same exposure to sunlight, light-skinned indi-

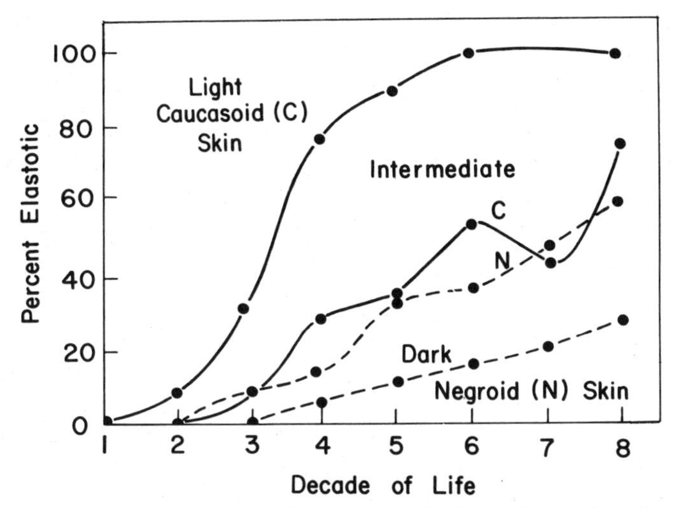

FIGURE 6:6. Relation between elastosis (the increase in elastic tissue and elastin) and decade of life for individuals with different degrees of skin pigmentation in a population mainly working outdoors in the northeastern United States. The percent of individuals with marked elastosis was determined by histological examination (orcein staining). Note that the dark Caucasoid (C) skin is only a degree more prone to elastosis than light Negroid (N) skin (intermediate curves). Data from Kligman, in *Sunlight and Man,* edited by T. B. Fitzpatrick *et al.,* Tokyo University Press, Tokyo, 1974, pp. 159 and 160.

viduals whose skins tan readily also retain more youthful skin longer than those who do not tan. Obviously, albino skin damages quickly. The skin of victims of xeroderma pigmentosum damages most rapidly. A single mild exposure may lead to marked skin damage because of lack of repair.

Aging

When normal skin that has been only mildly exposed to sunlight ages, it wrinkles but such wrinkling is shallow and not as prominent as in weather-beaten, furrowed, chronically sunburned skin. Aging skin is less resilient to stretch, and only slowly returns to its resting state.

Some aging dermal changes are similar to those with sunburn damage; others appear to be different. Elastic fibers in aging skin become less elastic much as in sun-damaged skin; they take up more salt (especially calcium), but do not undergo extensive fragmentation and other changes as in sun-damaged skin.

Although the decreased resilience of aged skin is superficially similar to that of sun-damaged skin, dermatologists report that changes in collagen fibers of aged skin are different from those of sun-damaged skin. Aging collagen fibers seem to develop increasing orderliness and crystallinity and become less soluble than in younger skin. They also develop greater resistance to the collagen-digesting enzyme, collagenase, and change chemically. These changes suggest more cross-linking between units and a greater degree of denaturation in the fiber proteins. In sun-damaged skin, the ground substance in which the fibers lie shows chemical alteration; in old age, similar phenomena occur.

Just as in sun-damaged skin, in aging skin the total number of sweat glands decreases and the remaining glands are not only less sustained in action but are slower, have a lesser total output, and are less responsive to drugs. In aging skin, as well as in sun-damaged skin, sebaceous glands become less active, resulting in a drier skin.

The thinning of the skin and its change in cross-sectional pattern and the disorganization of the layers of the epidermis seen in sun-damaged skin are not nearly as prominent in aging skin protected from the sun. A notable difference between young and old skin, neither one damaged by sunlight, is the capacity for reversal of sunburn damage in a transplant. This occurs in young but not in old skin. It suggests that even youthful-looking old skin has changed chemically.

For Additional Reading

Freeman, R. G. "Effects of Aging on the Skin." In *The Skin,* edited by E. E. Helwig and E. K. Mostopi. Baltimore: Williams and Wilkins, 1971, pp. 244–260.

*Kligman, A. M. "Solar Elastosis in Relation to Pigmentation." In *Sunlight and Man,* edited by Fitzpatrick *et al.* Tokyo: Tokyo University Press, 1974, pp. 157–163.

———. "Early Destructive Effects of Sunlight on Human Skin." *Journal of the American Medical Association* 210 (1969): 2377–2380.

*Lever, W. T. "Solar Elastosis." In *Histopathology of the Skin,* 4th ed. Philadelphia: Lippincott, 1967, pp. 265–286.

Mitchell, R. E. "Chronic Solar Dermatosis: a Light and Electron Microscope Study of the Dermis." *Journal of Investigative Dermatology* 48 (1967): 203–220.

*Montagna, W., ed. *Advances in the Biology of Skin,* vol. 6: *Aging.* Oxford: Pergamon Press, 1965.

———, Bentley, J. P., and Dobson, R. L. *Advances in the Biology of Skin,* vol. X: *The Dermis.* New York: Appleton-Century-Crofts, 1970.

———, and Hu, F., eds. *Advances in the Biology of Skin: The Pigmentary System.* Oxford: Pergamon Press, 1966.

———, Lobitz, W. C., Jr., eds. *The Epidermis.* New York: Academic Press, 1964.

Papa, C. M., Carter, D. M., and Kligman, A. M. "The Effect of Autotransplantation on the Progression or Reversibility of Aging in Human Skin." *Journal of Investigative Dermatology* 54 (1970): 200–212.

Pearse, R. H., and Grimmer, B. J. "Age and the Chemical Constitution of Normal Human Dermis." *Journal of Investigative Dermatology* 58 (1972): 347–361.

Summerlin, W. T., Charleton, E., and Karasek, M. "Transplantation of Organ Cultures of Adult Human Skin." *Journal of Investigative Dermatology* 55 (1970): 310–316.

7

Photosensitization of Cells

Biologists know that a cell is sensitive only to the radiation it absorbs, but that when the cell takes on a photosensitizer, it can absorb radiation it normally cannot. We know that cells of light-colored human skin normally absorb little visible and UV-A radiation, but all we need do is provide these cells with appropriate photosensitizers and they suddenly absorb these kinds of radiation. This can result in damage to the epidermal prickle cells, much like sunburn. All epidermal cells, whether in people, animals, or plants, and all free-swimming cells, are subject to photosensitization. By analogy, consider the photographic process, in which a pure silver chloride emulsion on film is sensitive only to the short end of the spectrum, the blue, violet and ultraviolet rays. If we add a dye, such as pinacyanol, it will absorb over the entire range of visible wavelengths. The dye acts as a sensitizer by transmitting the light energy from all the wavelengths it absorbs to the silver chloride, making the film sensitive to the entire rainbow of colors from violet to red seen by the human eye.

Photosensitizers occur in some plants, in smog, and in industrial wastes. They can reach the skin by direct contact. Human and animal epidermis may also be sensitized to light by ingesting contaminants as well as certain drugs and antibiotics. And, in addition, human and animal skin may be sensitized by products of cells with deranged metabolism in their bodies.

The number of substances applied to the skin externally (topical application) or taken internally (systemic administration) that serves as photosensitizers is more than trivial. Substances applied externally penetrate through the outer dead cells of the horny layer to reach the live cells in the prickle layer of the skin. Substances taken internally reach the live

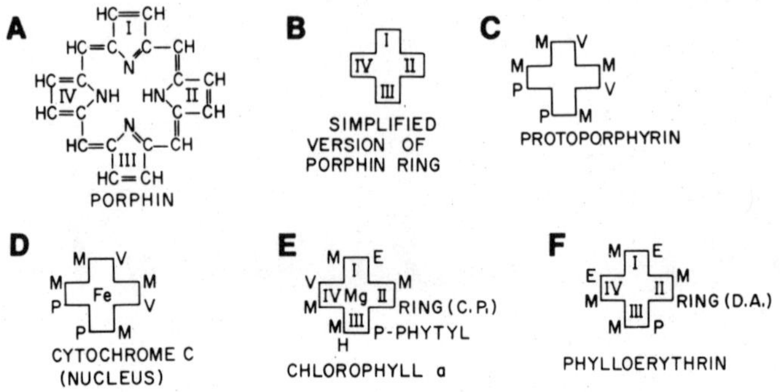

A PORPHIN

B SIMPLIFIED VERSION OF PORPHIN RING

C PROTOPORPHYRIN

D CYTOCHROME C (NUCLEUS)

E CHLOROPHYLL *a*

F PHYLLOERYTHRIN

LEGEND

C carbon atom
H hydrogen atom
N nitrogen atom
M methyl group, (CH_3)
E ethyl group, (CH_2CH_3)
V vinyl group, $(CH{=}CH_2)$
P propionyl group, (CH_2CH_2COOH)
Fe iron atom
Mg magnesium atom
Phytyl a hydrocarbon, $(C_{20}H_{39})$
Ring CP cyclopentanone ring
Ring DA same ring without acetyl group

FIGURE 7:1. Basic chemical structures of several porphyrins serving as cellular photosensitizers. (A) The porphin nucleus is common to all porphyrins. Note that it is made up of four conjoined but similar rings, each with a nitrogen atom attached to four carbons. This unit ring is called the pyrrole; a porphin is therefore a tetrapyrrole. (B) A simplified version of the porphin ring used in the subsequent diagrams. (C) Protoporphyrin, one of the natural porphyrins of widespread distribution. A porphyrin is a porphin with various atoms or groups of atoms substituted for outer hydrogen atoms of the original molecule. The legend below the figure describes the various substituents. Protoporphyrin is a forerunner of the natural cell pigment, cytochrome, and of the blood pigment, heme. (D) Cytochrome *c* nucleus, present in all aerobic (oxygen-burning) cells. Note that it has an iron atom at the center of the nucleus; it is therefore called a metalloporphyrin. The heme of hemoglobin in our blood has the same structure. (E) Chlorophyll *a*, another metalloporphyrin, the green pigment functioning in photosynthesis in plants. Note that the substituents for the outer hydrogens of the porphin nucleus are different from those in protoporphyrin; especially noteworthy are the hydrocarbon, phytyl, and the 5-carbon ring (CP, for cyclopentanone), which are all characteristic of chlorophyll. Chlorophyll is the molecule in the photosynthetic plant that absorbs the light energy necessary for fixation of the carbon into carbohydrates, etc. (F) Phylloerythrin, an altered chlorophyll molecule produced by bacterial action in the digestive organs of cud-chewing vertebrates. It can be absorbed from the gut into the bloodstream largely because it has lost the large phytyl hydrocarbon chain that normally prevents entry of chlorophyll into cells lining the gut.

cells through blood and tissue fluid. Upon illumination, photosensitization is revealed in cells of the prickle cell layer.

Natural Photosensitizers

Some natural photosensitizers have been known for thousands of years; others, only in modern times. Their mechanism of action has been studied only recently. While the main groups of natural photosensitizers—the porphyrins, hypericins, and furocoumarins—have been under considerable study, the others are less well known.

Porphyrins. Porphyrins, found in all cells, comprise a family of compounds with a complex chemical structure. The members of this family differ from one another by substituents attached to the ring, as indicated in Figure 7:1A to F. As seen in Figure 7:1D, the respiratory pigment cytochrome, *c,* found in all aerobic cells, contains an iron atom in the center of the molecule. A similar porphyrin, called heme, attaches to the protein globin to form our blood pigment hemoglobin. In chlorophyll, shown by Figure 7:1E, magnesium is present in the center of the porphyrin molecule and is further distinguished by two unique substituents, a 5-carbon ring (cyclopentanone) and a long-chain hydrocarbon (phytyl).

Apart from chemical interest in the precise arrangement of the atoms of porphyrins, most important for our story is the fact that they absorb short wavelengths of visible light and UV-A radiation. This property gives them their photosensitizing power when absorbed radiant energy passes to vulnerable cell structures. For example, hematoporphyrin, produced in the laboratory from natural porphyrins, sensitizes colorless cells such as sea urchin eggs or paramecia to visible and UV-A radiation causing cell disruption and death. In the dark the porphyrin has no effect, and colorless cells are not affected by the same dose of light in the absence of the sensitizer. Porphyrin-sensitized reactions are photooxidations and occur only in the presence of oxygen.

In 1912, the German physician, Meyer Betz, sensitized himself to visible light by intravenously injecting into his arm a small amount of dissolved hematoporphyrin. He became extraordinarily photosensitive even to artificial light and, as the photographs in Figure 7:2 show, his face swelled. His breathing became labored as well. The effects lasted for

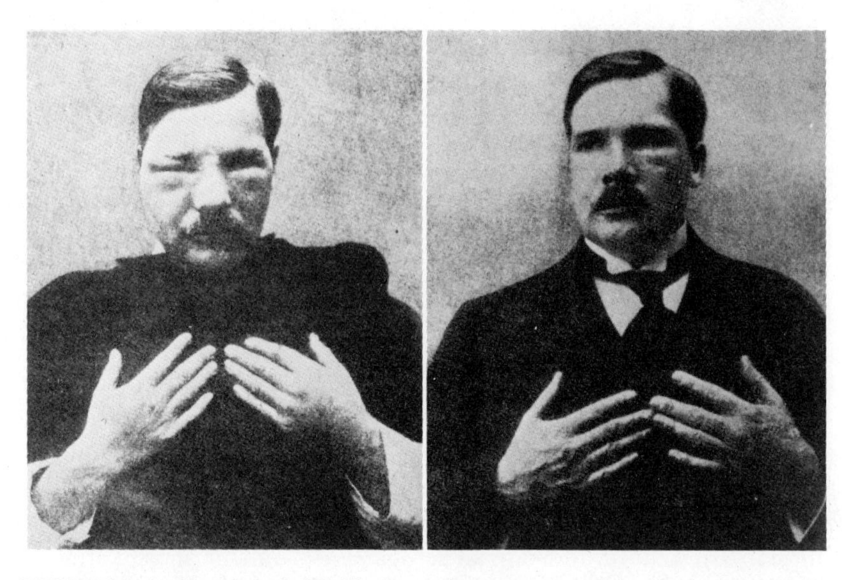

FIGURE 7:2. Swelling of the facial skin observed on exposure to light after intravenous injection of 0.2 g hematoporphyrin. Hematoporphyrin is like protoporphyrin (Figure 7:1C) but with hydroxyethyl ($CHOHCH_3$) groups in place of the vinyl groups (V) present in protoporphyrin. Dr. Meyer Betz, who performed this experiment on himself ("Selbstversuch") on October 14, 1912, was still sensitive to light at the beginning of December, 1912. By springtime he was completely desensitized. From Laurens, *The Physiological Effects of Radiant Energy,* Chemical Catalog Co. (Tudor), New York, 1933, p. 496.

about three months, then gradually lessened until, fortunately, they were completely gone. Hematoporphyrin has not been found naturally in man and in other animals, but natural porphyrins, produced in excess in deranged metabolism, may give similar effects.

Porphyrias. In normal human and stock animal metabolism, certain porphyrins are synthesized as needed. The rate of synthesis is regulated by cellular mechanisms. When cells in the body fail to produce an adequate supply of a key enzyme in the porphyrin synthetic process, a manufacturing bottleneck causes the precursor molecules feeding into this step to accumulate. For example, when the body becomes deficient in the enzyme catalyzing the binding of iron into protoporphyrin to form heme, which ultimately combines with the protein globin to form hemoglobin in our red blood cells, protoporphyrin accumulates, causing a disorder. In similar deficiencies of other enzymes of porphyrin metabolism, other precursors accumulate. All these disorders are collectively called porphyrias. When the accumulated precursors are completely voided in feces and urine (which may become burgundy red), they do not reach the skin. In cattle, in the disorder known as "pink tooth disease," excess porphyrin accumulates in the bones and teeth. When the precursor enters the bloodstream and reaches the prickle cells, the skin is sensitized to light.

The general types of hereditary porphyrias appear either early or late in life. Other porphyrias, induced by poisons, drugs, or disease, may also occur upon exposure either early or late in life.

Physicians believe that a number of drugs induce porphyrias in some people. These include hypnotics, sedatives, anticonvulsants, analgesics, and antirheumatics. Other chemicals such as solubilizers, fungicides, and insecticides are also likely to cause porphyrias. When porphyrins invade the skin, they sensitize it to light. Ethyl alcohol, some fungicides, normal estrogens regulating the monthly estrus cycle (as well as the synthetic estrogen stilbestrol), and the antimalarial compound chloroquin are also believed to aggravate porphyrias.

Those with sensitizing porphyrin in their skin suffer from a marked stinging and burning of the skin in sunlight, even when filtered through a window or passed through thin clothing. As in sunburn, the victim develops erythema, blistering, scabbing, and scaling or peeling. The effects of a single exposure may last a few hours, a few days, or, in extreme cases, several weeks.

In the disease known as protoporphyria, protoporphyrin (depicted in Figure 7:1C), the precursor of the heme part of hemoglobin, accumulates in red blood cells. When removed from a blood vessel, they show marked sensitivity to light. Little light of the wavelengths absorbed by protoporphyrin reaches blood cells through the skin. When protoporphyrin accumulates in the skin, however, the skin cells show photosensitivity.

Curiously, porphyrins show no photosensitizing action when applied to human or mouse skin. Even when the horny epidermal layer is stripped from the skin with adhesive tape or the skin is scratched and porphyrin is applied, the skin cells are not sensitized to light. Even in experiments where prophyrin is fed to mice, the skin cells are not sensitized. To produce photosensitivity, porphyrin must be injected into the skin, body fluid, or blood. While no one has satisfactorily explained why porphyrin works as it does, it is possible that the pigment reaches epidermal prickle cells only through the bloodstream.

Porphyrias also occur in single-celled organisms. Some bacteria, for example, strains of *Myxococcus xanthus,* produce more porphyrin than they need, and it accumulates in the medium when the bacteria no longer divide. If illuminated in the presence of oxygen, the bacteria are killed, destroyed by the very prophyrin photosensitizers they produced. A strain of the ciliate protozoan, *Tetrahymena,* with a block in its porphyrin synthetic metabolism, similarly destroys itself.

Chlorophyll. As a porphyrin, chlorophyll also acts as a photosensitizer in cells, but only under certain conditions. For example, when the blue-green mutant of the purple photosynthetic bacterium *Rhodopseudomonas* is illuminated in the presence of oxygen, not only is chlorophyll synthesis stopped but bacterial growth is also inhibited. In experiments where the illuminated bacteria are spread on a jellied culture medium, investigators find that colonies cannot form, indicating that the cells cannot divide. As seen in Figure 7:3B, illumination of a similar culture of the blue-green mutant in the *absence* of oxygen has no effect on either chlorophyll synthesis or viability. The action spectrum shows that the light receptor inducing these effects is chlorophyll. While the wild-type, purple-colored form of the same species shows cessation of chlorophyll synthesis under illumination in the presence of oxygen, it is protected from photosensitized growth inhibition by its purple carotenoid pigments that serve as antioxidants. The blue-green variety has an unprotective colorless carotenoid, lacking the reducing power of the purple type. In experiments with the purple-colored, wild type where the synthesis of protective carotenoids was inhibited by the drug diphenylamine, the wild variety became sensitive to light in the presence of oxygen, in very much the same way as the blue-green mutants. In other experiments, investigators have found that carotenoids with more than seven double bonds, as found in the purple carotenoids, protect dissolved chlorophyll from photooxidative destruction by oxygen activated with the energy absorbed by the chlorophyll, as is also the case of other porphyrins.

Similarly, biologists have discovered photosensitivity in some mutant green algae and in some mutant sunflower and Indian corn plants. Light sensitivity in these cases was found to correlate with the absence of

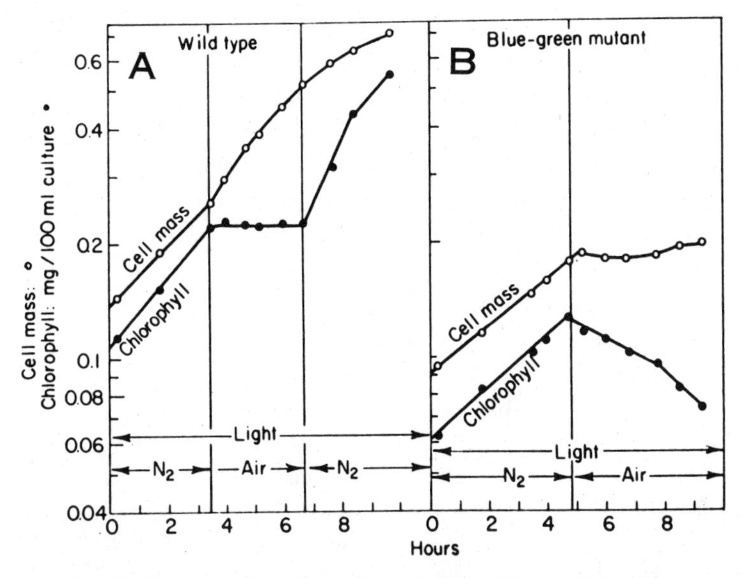

FIGURE 7:3. Inhibition of chlorophyll synthesis on illumination in the presence of oxygen in the purple photosynthetic bacterium *Rhodopseudomonas spheroides,* in the wild type and in the blue-green mutant. In the wild type (A) chlorophyll synthesis merely ceases (level portion of chlorophyll curve) during illumination in air, but in the mutant (B) illumination in air is accompanied by killing of the bacteria as indicated by inoculation on growth media (not shown). From Sistrom *et al., Journal of Cellular and Comparative Physiology* 48, 1956, p. 473.

carotenoids with a large number of double bonds. Interestingly, carotenoids (β-carotene) are protective in human porphyrin photosensitization. When stock animals or humans eat chlorophyll from plants, it does not penetrate the gut wall and so cannot enter the bloodstream and be carried to the skin to act as a sensitizer.

Yellow Thick Head. Witnessed primarily in sheep in South Africa, this photosensitivity disease is induced by a porphyrin formed during chlorophyll digestion by bacteria living in the sheep gut. During digestion of plant food, the hydrocarbon (phytyl) side chain of chlorophyll and an acetic acid residue (in the 5-carbon cyclopentanone) ring are removed, converting chlorophyll to phylloerythrin (as shown in Figure 7:1F), which, unlike chlorophyll, penetrates the gut, goes through the blood, and then is carried to the bile. Normally, the phylloerythrin that reaches the intestine by way of the bile is voided in the feces. However, when the sheep eat particular wilting plants that contain a chemical that blocks the bile duct, the bile builds up because it cannot be discharged. The phyllo-erythrin still produced in the same way as in normal sheep is then reabsorbed into the bloodstream and through the blood reaches the skin, sensitizing it to light. Ultimately, this series of events leads to swelling and edema, whence the name "thick head." Bile-duct blockage also causes uptake into the blood of the bile salts, some of which are deposited in the skin, whence the yellow color of "yellow thick head" (the Africaans *geeldikkop*).

Yellow thick head disorder can be artificially produced by tying off (ligating) the bile duct and feeding greens (containing chlorophyll) to the sheep. If grain is fed instead of greens, the skin of the jaundiced sheep gets yellow, but in the absence of chlorophyll in the diet, light sensitivity does not develop and the "thick head" disorder does not appear. A similar disorder has also been seen in some herbivores other than sheep.

Hypericins. Hypericins are red-colored organic compounds found in glands on the surface of some species of the plant genus *Hypericum,* drawn in Figure 7:4. The western range plant *Hypericum perforatum,* or St. Johns wort, is best known in the United States, but the genus *Hypericum* is large, with close to three hundred species all over the world. Of these, about two-thirds produce the compound hypericin, which photosensitizes the skin. Extracted from a European species, hypericin was purified and its structural formula, given in Figure 7:5, was determined and then it was synthesized.

In regions where *Hypericum* grows, white animals are impossible to keep, particularly white horses in parts of Arabia and white sheep in some regions in Italy. After white animals eat these plants and are exposed to light, they develop itchiness and swelling of the skin, especially about the mouth, nose, eyes, ears, and hooves, and sometimes over their entire bodies. The disorder is called hypericism. While white animals are kept in darkness after eating the plants, no photosensitization develops. If such animals are fed grain only, they show none of the sensitivity reactions even when kept in the sun.

Hypericism, once a threat to range lands in Australia and the United States, fortunately no longer darkens the economic horizon. The offending plants were eradicated by a successful method of biological control using the beetle *Chrysolina gemellata,* which feeds only on plants of the genus *Hypericum.* Its life cycle is well coordinated with growth of the plant.

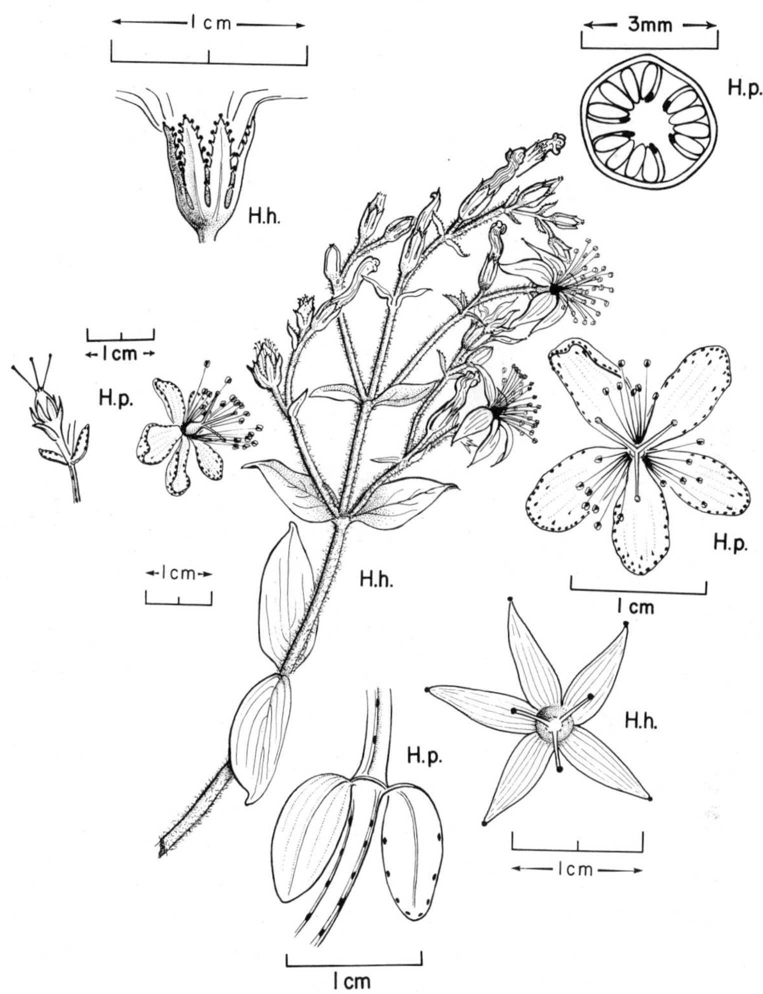

FIGURE 7:4. Two species of the genus *Hypericum,* showing as black dots the glands that contain the photosensitizing pigment hypericin: *Hypericum perforatum* (H. p., St. John's wort or Klamath weed), introduced into North American and Australian range lands, which has glands on the petals (middle right and left), anthers (middle right) and within the flower ovary, between each row of developing seeds (shown in section at top right), and on stem ridges and the undersurface of leaves (lower middle); *Hypericum hirsutum* (H.h.), of Europe, used in chemical characterization of the pigment, with pigment glands on only the tips of floral stigmas and petals (lower right) and sepals (upper left).

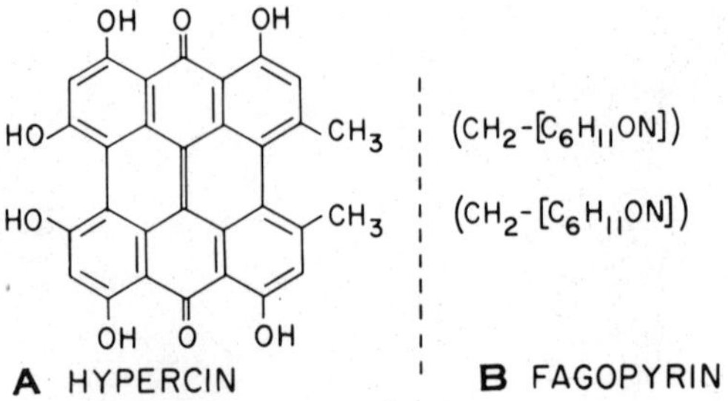

OH O OH
HO CH$_3$ (CH$_2$-[C$_6$H$_{11}$ON])
HO CH$_3$ (CH$_2$-[C$_6$H$_{11}$ON])
OH O OH
A HYPERCIN **B** FAGOPYRIN

FIGURE 7:5. (A) Structure of hypericin, a photosensitizer found in many species of the genus *Hypericum,* including *H. hirsutum* in Europe (Figure 7:4) and *H. perforatum,* common to our range lands. (B) In fagopyrin, found in buckwheat, the groups shown in parentheses are substituted for the methyl (CH_3) groups of hypericin.

A form of hypericism, called fagopyrism, is attributed to buckwheat (*Fagopyrum esculentum*). The main photosensitizing agent present on the green parts of this plant is a pigment structurally like hypericin except for substitutions for each of two methyl groups in hypericin (see Figure 7:5). Buckwheat "poisoning" was brought under control by not feeding the greens to domestic animals. Only the seeds, which lack hypericin, are now harvested and used. Cases of fagopyrism still occur occasionally when animals at pasture wander into fields containing buckwheat plants.

In the ciliate protozoan *Blepharisma,* a pink hypericinlike pigment is found in small granules at the outer surface of the cell (see Figure 7:6). The pigment sensitizes *Blepharisma* itself to bright light in the presence of oxygen, resulting in a large increase in oxygen consumption and death of the cell. Placed in dim light, the pink pigment of *Blepharisma* undergoes slow photooxidation to blue-gray, in which state it does not act as a damaging photosensitizer. Experiments with extracts of both pigments show that the red pigment sensitizes cells to light, while the blue-gray pigment does not. In nature, *Blepharisma* is a bottom dweller and so avoids exposure to bright light. In the laboratory, apparently by trial and error, it hides behind stones or detritus to avoid bright light. When unable to hide, it turns blue-gray in the early morning and becomes relatively resistant to bright light.

Hypericins sensitize human skin to light, but since it is very unlikely that people would naturally acquire the pigment either by contact or ingestion, hypericins do not present us with a problem.

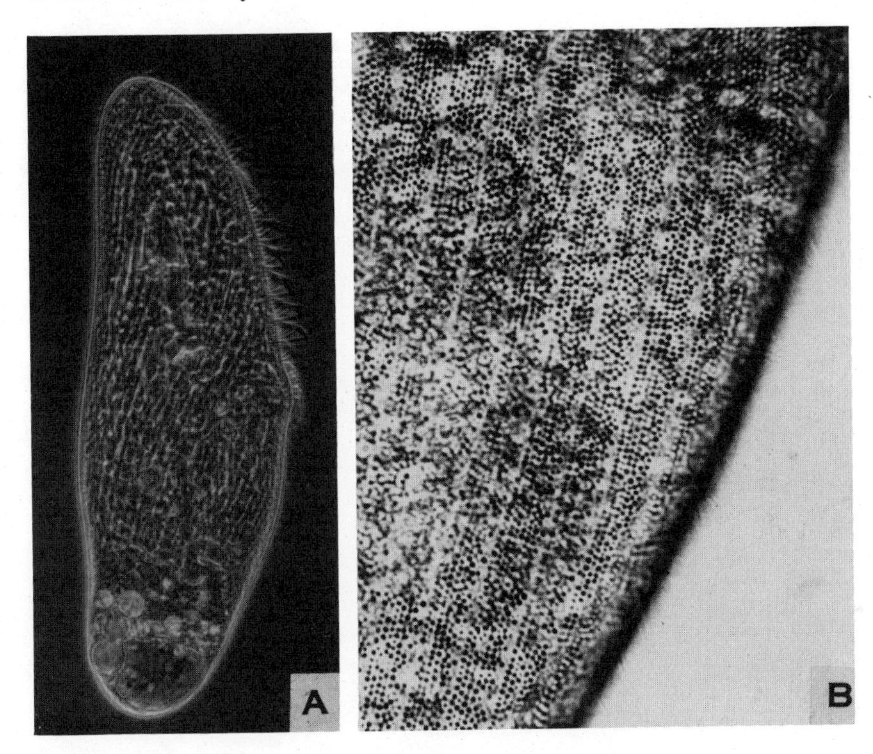

FIGURE 7:6. (A) The single-celled, pink, ciliate protozoan *Blepharisma japonicum* living in fresh water ponds. Ciliary rows run the full length of the cell from 200 to 300 μm (millionths of a meter) tip to stern. Between the ciliary rows are arrays of pigment granules shown in the photomicrograph (B) taken under oil immersion.

All hypericins are photodynamic dyes and are photosensitizers only in the presence of oxygen. In some manner the activation energy of the light absorbed by the dye is passed to oxygen which oxidizes the surface of cells, disrupting the cell membranes.

Furocoumarins. The furocoumarins are three-ringed organic compounds found in a variety of plants, as shown in Table 7:1. Popularly called psoralens, they are composed of conjoined coumarin and furan rings, as depicted in Figure 7:7. Only certain of the furocoumarins are effective photosensitizers. Alone, neither furans nor coumarins are photosensitizers.

All furocoumarins have the same general three-ringed nucleus, but differ from one another in the kinds of substitutions in various positions

Table 7:1. Families of Plants Causing Contact Photosensitization[a]

Family	Common name	Example (species)
Furocoumarins identified in extracts		
Umbelliferae	Parsley family	Garden parsley (*Petroselium sativum*)
		Angelica (*Angelica archangelica*)
		Bishop's weed (*Ammi majus*)
Rutaceae	Rue family	Bergamot (*Citrus bergamea*)
		Rue (*Ruta graveoleus*)
		Lemon (*Citrus limonum*)
Moraceae	Fig family	Fig (*Ficus carica*)
Leguminosae	Legume	Bavachi (*Psoralea corylifera*)
Furocoumarins suspected		
Ranunculaceae	Buttercup family	Buttercup (*Ranunculus* sp.)
Cruciferae	Mustard family	Mustard (*Brassica* sp.)
Convolvulaceae	Morning glory family	Bindweed (*Convolvulus arerensis*)
Rosaceae	Rose family	Agrimony (*Agrimonia eopatoria*)
Compositae	Sunflower family	Yarrow (*Achillea millifolium*)
Chenopodiaceae	Saltbrush family	Goosefoot (*Chenopopium* sp.)

[a] After Giese, *Photophysiology,* Academic Press, New York, Vol. VI, 1971, p. 113.

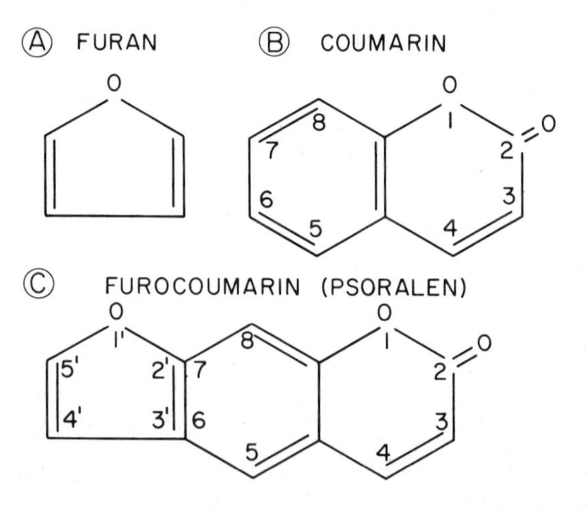

FIGURE 7:7. A furocoumarin (C), which is a photosensitizer, is formed by combining furan (A) and coumarin (B) rings, neither of which separately has photosensitizing action on the skin cells or cell suspensions.

on the rings. Many have been isolated and studied, and quite a few have been made synthetically. It was found that while a few kinds of substitutions increase the photosensitizing potency of the compound, many others decrease photosensitizing potency, and some destroy it completely.

Long before their chemical nature was known, some furocoumarins were used to treat leucoderma (meaning white skin). Thousands of years ago in India, they applied an extract of the black seeds of the legume *Psoralea corylifera* to white spots on the skin and then exposed the skin to light. Much as in sunburn, erythema and tanning followed and with periodic treatments the skin got darker and eventually matched the surrounding skin. Similar treatment was used in the ancient Nile Valley and later in China. Interestingly, dermatologists use the same therapy, but now the substance may be administered orally. Sometimes sunbathers use furocoumarins to quicken tanning.

Present in minute superficial glands on the surface of the flowers, stems, leaves, and seeds of some plants, furocoumarins have been extracted from plants in four families. While plants in several other families induce skin irritation, furocoumarins have not yet been extracted from them. Table 7:1 indicates the plants in which furocoumarins have either been identified or are suspected to occur.

When furocoumarins and related compounds are applied to the skin, then illuminated, they induce a skin condition known as photophytodermatitis, or plant-photosensitivity skin disease. In nature, the condition results from exposure to light after either direct contact with the plants or contact with some object like a towel recently in contact with the plants.

In many ways, photosensitization by furocoumarins resembles sunburn, except that it is induced by relatively nonsunburning wavelengths of light in the blue violet and UV-A portion of the spectrum (between 440 and 334 nm). Erythema develops within one to twenty-four hours after exposure. The degree of redness and other symptoms depend on the amount of the sensitizer applied and the intensity and duration of the light. As in sunburn, lysosomes in the prickle cells rupture, and the cells die. When many prickle cells are killed by photosensitized action, the skin may blister as in severe sunburn. Later, the dead layers peel off and a tan develops. Furocoumarin-photosensitized tanning in light-skinned people is accomplished by transfer of larger melanosomes than in sunburn tanning. Furthermore, the melanosomes do not fragment but retain their identity as in dark-skinned people (see Figure 5:6D). Tanning after photosensitization is more persistent than after sunburn.

In testing a variety of furocoumarins on bacteria, scientists found that only the furocoumarins that photosensitized human skin killed bacteria and oxygen was not required. In these studies, interestingly enough, biologists discovered many mutants among the survivors of the photosensitized bacteria, suggesting that furocoumarin photosensitization was mutagenic.

Since mutation is a consequence of DNA injury, research workers tested the possible photosensitization of DNA by furocoumarins. They found that DNA forms a complex with the furocoumarin in light but not in darkness. The complex had properties different from those of pure DNA. Applied to the skin, furocoumarins formed a complex with the DNA in skin cells, but not with the skin proteins. Experiments demonstrated that furocoumarins reacted with the pyrimidine base thymine in DNA to form an addition product. Figure 7:8 shows also that in solution and following illumination, furocoumarins form addition products with thymine. Recent studies reveal that some furocoumarins, such as psoralen, can bind two thymine molecules, one on each side. When reacting in this manner with DNA, the psoralen can cross-link the two strands of a DNA molecule, intercalating itself between the two. The addition products can be split in dark repair.

Both types of addition products inhibit DNA replication. The cell removes such replication-inhibiting damage by the cut and patch dark recovery mechanism described in Chapter 4. No photoreactivation of such damage has been observed, nor is it likely to occur in view of the specificity of the photoenzyme for pyrimidine base dimers.

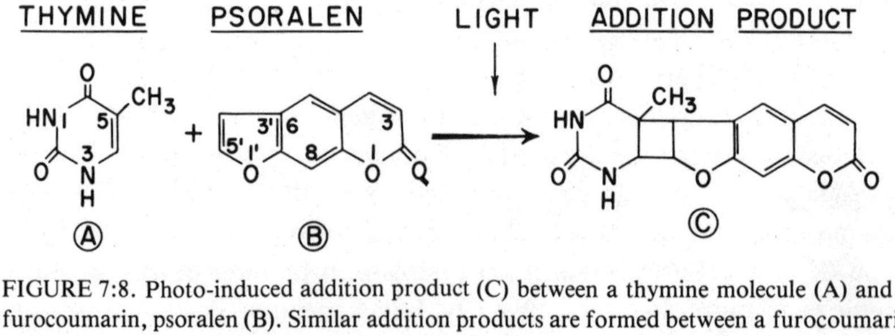

THYMINE PSORALEN LIGHT ADDITION PRODUCT

FIGURE 7:8. Photo-induced addition product (C) between a thymine molecule (A) and a furocoumarin, psoralen (B). Similar addition products are formed between a furocoumarin molecule and a thymine component in a DNA molecule.

In RNA, furocoumarins form an addition product with a uracil base. No one knows whether the reaction of furocoumarins with RNA in the cell is of much consequence to the viability of the cell, because there is so much more RNA in the cell compared to DNA.

Studies with furocoumarins demonstrate that photosensitization can occur by action on cellular DNA, in contrast to action of porphyrins and hypericins primarily on cell membranes. Because furocoumarins may induce mutations, such mutagenic action on body cells of organisms coming in contact with plants bearing furocoumarins might result in cancerous growths. However, no evidence to check this possibility in humans or animals is available. The way in which drugs or atmospheric pollutants discussed below cause photosensitization is not known. They may act on cell surfaces like porphyrins and hypericins or on DNA like furocoumarins, or in still another manner. (Recent studies show that some porphyrins may react with DNA.) However, the possibility of action on DNA must be borne in mind, given the mutagenic hazard it involves.

As late as 1974, furocoumarin sensitization of the skin to light has been put to effective use in controlling psoriasis. In this disease both DNA synthesis and cell division in the basal cells of the epidermis accelerate, and maturation of the basal cells to squamous cells takes only three to four days instead of the usual twenty-eight days in normal skin. What induces the loss in body control over DNA synthesis and rate of cell division in psoriasis is not known, but the disorder is widespread and is said to affect about one to three percent of the world's population. Excessive cellular proliferation forms silvery epidermal placques that scale off with intense itching, but it does not result in cancerous growths. Psoriasis sufferers are treated with an oral dose of a furocoumarin (8-methoxypsoralen). Two hours afterward, when maximum skin concentration has been reached, the patient is exposed to radiation from a bank of fluorescent lamps designed to emit an intense band of UV-A radiation, including the peak of absorption by the furocoumarin. A series of such skin treatments, during which DNA synthesis and cell division in the basal cells of the epidermis are inhibited, leads to a complete remission of the symptoms. The skin becomes essentially normal, except for the development of a deep tan. With this treatment, the skin stays normal for a longer period of time than with other therapies. Occasional periodic treatments maintain normalcy. Damaging effects of furocoumarin photosensitization are not anticipated, because white spots on the skin (vitiligo) have been treated in this manner for some twenty years without signs of cancer. Clinicians will be testing this therapy for sometime before it becomes standard, just in case some as yet unknown secondary effects appear.

Herpes virus is inactivated by light after photosensitization with photodynamic dyes. This may permit control of the painful venereal form of the disease. Successful treatment of mouse tumors by dye photosensitization to deeply penetrating laser beam light gives promise of application to superficial human tumors.

Photosensitization by Medications

Some types of medication produce photosensitization of two types, phototoxic and photoallergic. The former occurs on first exposure to light in the presence of the sensitizer, and the latter only after a lapse of time on a second challenge with light and photosensitizer.

Phototoxic Reactions. Photosensitization of skin by industrial effluents, smog, and medications provokes reactions similar to natural photosensitizers. Substances eliciting phototoxic reactions belong to a large number of chemical families, which are classified here on the basis of how they are used in medicine: the B vitamin riboflavin; the hormones estrone (female sex hormone) and diethylstilbestrol, used in treating menopausic symptoms and suppressed lactation; some antibiotics; the antimalarial quinine; the antisyphilitic, arsphenamin; the antileukemic drug, triethylene melamine; the diuretic, chlorothiazide; the cyanide and nitrate poison antidote, methylene blue; the tranquilizer, barbiturate; some surfactants and antibacterials in soaps, detergents, and cosmetics; dyes, in lipsticks; psoralens, in perfumes, and various coal tar products (pyridine, acridine). The list is not complete, but illustrates the widespread occurrence of photosensitizing materials. Phototoxic photosensitization occurs only while the substances are present in the skin.

How photosensitization is induced by these substances is now being studied by dermatologists. Usually, UV-A radiation and visible light are effective and, where action spectra have been determined, they approximate the absorption of these radiations by the compounds involved.

Not everyone is photosensitized by these preparations. Much depends on the amount of melanin in the epidermis. Dermatologists have found that phototoxic reactions are uncommon in dark-skinned people in whom sufficient energy of effective wavelengths of light does not reach the prickle cells of the epidermis. Other factors may also play a role.

Photoallergic Reactions. In a photoallergic reaction, light converts a photosensitizer into an allergen, a substance which evokes the formation of an antibody specific to it. An allergen may act by itself or alternatively combine with a protein and serve only as the active group (haptene), giving specificity to the antibody formed against the substance. The antibody is formed by cells in the body of the challenged animal. After an incubation period during which these events transpire, another challenge results in an inflammatory antigen–antibody reaction. Table 7:2 summarizes some of the clearer differences between phototoxic and photoallergic reactions. Both dermis and the epidermis may be involved in antigen–antibody reactions.

As an example of a photoallergic reaction, consider what happens to some people who take the tranquilizer chlorpromazine. When exposed to light for a second time, they develop an inflammatory reaction. They suffer from a watery discharge, followed by scales and crusts, accom-

Table 7:2. Characteristics of Phototoxicity and Photoallergy[a]

Reaction	Phototoxic	Photoallergy
On first exposure	Yes	No
Incubation period necessary after first exposure	No	Yes
Photochemical alteration of photosensitizer	No	Yes
Observed response in the skin	Usually resembles sunburn	Varied: erythema, edema, vesiculation etc.
Concentration of reagent required	High	Low
Passive transfer from one individual to another	No	Possible
Action spectrum	Similar to absorption spectrum	Usually longer wavelengths

[a]Adapted from Harber, "Pathogenic Mechanism of Drug-Induced Photosensitivity," *Journal of Investigative Dermatology* 58 (1972), p. 327.

panied by intense itching and a burning sensation (eczema). In solution, exposed to light, and then applied to the skin in the dark, chlorpromazine provokes the same reactions. After taking the drug, if the individual stays in the dark, or if the unilluminated drug is applied to the skin in darkness, there is no reaction. Dermatologists believe that what happens is that the chlorpromazine gets to the skin through the bloodstream. When light hits the skin, it is photochemically converted into an allergen; the same allergen is developed on illumination of the drug in solution. The allergen provokes the creation of antibodies, and with a second challenge, an inflammatory reaction results. Figure 7:9 presents a diagram of

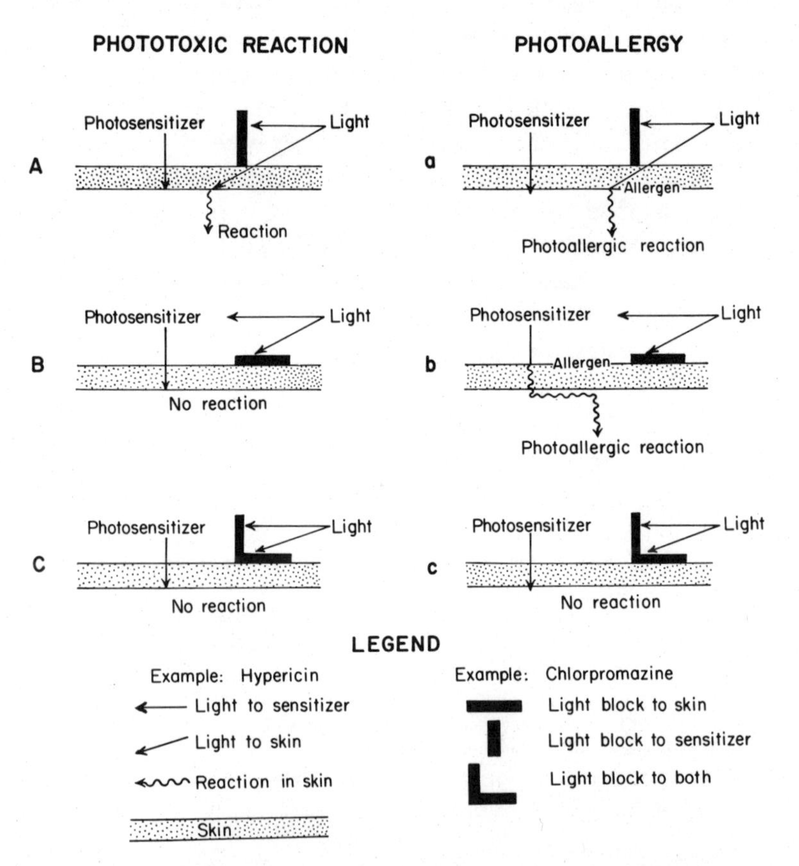

FIGURE 7:9. Diagram showing the difference between phototoxic and photoallergic reactions of the skin to light in the presence of photosensitizers. Modified after Ippen, in *The Biologic Effect of Ultraviolet Radiation,* edited by F. Urbach, Pergamon Press, Oxford, 1969, p. 517.

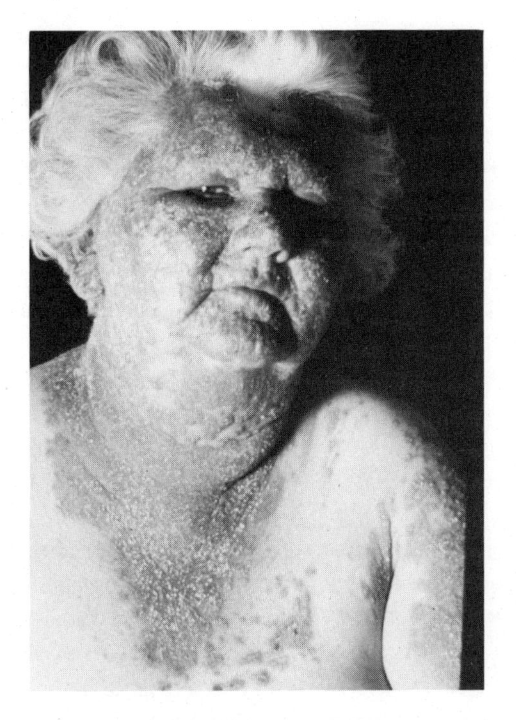

FIGURE 7:10. A photoallergy provoked by exposure to light after treatment with sulfathiazol. Courtesy of Dr. H. Ippen, University of Göttingen.

these events, while Figure 7:10 shows what can happen to someone afflicted with an allergic skin reaction.

The major photoallergenic medications are some antibacterials, (sulfanilamides, for example), antifungals (Griseofulvin is one), antihistamines (like Phenergan—promethazine—in cough medicine), cyclamates, diuretics (thiozides), oral hypoglycemic agents (Orinase), some sun screens (substituted benzoic acids), and tranquilizers (chlorpromazine).

Not everyone develops photoallergic reactions to these drugs. Those who are prone to allergies are usually affected, and substances that are only phototoxic to some may be photoallergenic to others.

Some pathological conditions also result in photoallergies upon illumination of the skin. Damage may occur to blood vessels, connective tissue, and even to internal organs such as the liver. The photoallergens in question are produced by body cells (endogenous) and have not yet been identified.

Photosensitization in Diseased Conditions

At the turn of the century, Danish physician and pioneer phototherapist Niels Finsen observed that daylight irritated the pustules of smallpox. He recommended that patients be kept in a very dimly illuminated room or in red light. This technique, called negative light therapy, is also used on patients who suffer from viral infections, such as lymphogranuloma and vaccinia vaccination, that also induce phototoxic reactions. Neither the photosensitizer nor the action spectrum for the effect is known. Nevertheless, as with Finsen's patients, those afflicted must be protected from bright light.

Protection from Photosensitization

Since photosensitization occurs chiefly because of blue-violet and UV-A radiations, one can protect oneself by using chemical screens that remove these wavelengths. Screens can either be taken orally, as in the case of β-carotenes used for porphyrias, or applied to the skin as lotions or creams. Sometimes, it is necessary to avoid daylight or bright illumination altogether. If, on occasion, the patient must go out into the light, the skin must be covered with light-impervious clothing and gloves. Naturally, the best protection is to avoid contact with photosensitizers whenever possible.

For Additional Reading

Blum, R. F., *Photodynamic Action and Diseases Caused by Light,* Reinhold, 1964.

Diamond, I., McDonaugh, A. F., Wilson, C. B., Gronelli, S. G., Nielson, S., and Joenicke, R. "Photodynamic Therapy of Malignant Tumors." *Lancet* 2 (1972): 1175–1177.

Epstein, J. H., and Fukuyama, K. "Effects of 8-MOP-Induced Phototoxic Effects of Mammalian Epidermal Macromolecular Synthesis *in Vivo.*" *Photochemistry and Photobiology* 21 (1975): 325–330.

Fitzpatrick, T. B., Pathak, M. A., Harber, L. C., Seiji, M. and Kukita, A. eds. *Sunlight and Man,* Tokyo: Tokyo University Press, 1974, pp. 314–333, 335–368, 359–387, 459–477, 495–513, 631–653, 659–668, 783–791.

Giese, A. C. "Photosensitization by Natural Pigments." In *Photophysiology,* vol. 6, edited by A. C. Giese. New York: Academic Press, 1971, pp. 77–129.

———. *Blepharisma—The Biology of a Light-Sensitive Protozoan,* Stanford, California: Stanford University Press, 1973.

Grossweiner, L. I. "Molecular Mechanisms in Photodynamic Action." *Photochemistry and Photobiology* 10 (1969): 183–189.

Harber, L. C., and Baer, R. L. "Pathogenic Mechanisms of Drug-Induced Photosensitivity." *Archives of Dermatology* 58 (1972): pp. 327–342.

Jarratt, M. "Phototherapy of Dye-Sensitized Herpes Virus." Abstracts Third Annual Meeting, American Society for Photobiology, (1975): pp. 40–42.

Krinsky, N. I. "The Protective Function of Carotenoid Pigments." In *Photophysiology,* vol. 3, edited by A. C. Giese. New York: Academic Press, 1968, pp. 123–195.

Musajo, L. and Rodighiero, G. "Mode of Photosensitizing Action of Furocoumarins." In *Photophysiology,* vol. 7, edited by A. C. Giese. New York: Academic Press, (1972), pp. 115–147.

Parrish, J. A., Fitzpatrick, T. B., Tanenbaum, L., and Pathak, M. A. "Photochemotherapy of Psoriasis with Oral Methoxalen and Longwave Ultraviolet Light." *New England Journal of Medicine* 291 (1974): 1207–1211.

Phototherapy of the Newborn: an Overview. Washington D.C.: National Academy of Science USA, 1975.

Runge, W. J. "Photosensitivity in Porphyria." In *Photophysiology,* vol. 7, edited by A. C. Giese. New York: Academic Press, 1972, pp. 149–162.

Scheel, L. O. *Toxicants Occurring Naturally in Food Materials,* edited by F. M. Strong. Washington, D.C.: National Research Council, National Academy of Science, U.S.A. 1973, pp. 558–572.

Shelley, W. B. "Photosensitizers." In *Dermatoses Due to Environmental and Physical Factors,* edited by R. B. Rees. Springfield, Illinois: Thomas, 1962, pp. 88–103.

*Urbach, R. ed. *The Biologic Effects of Ultraviolet Radiation.* Oxford: Pergamon Press, 1969, pp. 489–511, 513–525, 527–532.

8

Sunlight and Cancer

The most common human cancer, skin cancer, has long been known to be localized mainly on the head and neck. Since these areas are most directly exposed to the sun, it is no surprise that sunlight was suspected as the culprit. Physicians keep two kinds of records on human skin cancer: incidence and prevalence. Incidence data record the number of cases of cancer appearing in a given geographical area per year, and prevalence data indicate the total number of cancers at one time in a given place. Squamous cell carcinoma and basal cell carcinoma, shown in Figures 8:1B,C, are the two most commonly found types of skin cancer. These carcinomas may form nodules on the skin or grow downward into the dermal connective tissue, or invade the dermis laterally. Skin cancers appear to be on the increase in the light-skinned population.

The squamous cell carcinoma originates in the partially differentiated cells of the prickle cell layer (see Figure 5:2P). Reports reveal about twenty-five cases per one hundred thousand people per year at 45°N latitude, doubling with each 8 to 11° decrease in latitude. These cells undergo partial differentiation forming keratin and are always identifiable as squamous cells. They multiply without control after presumed chronic injury of the cells by sunlight (see Figure 8:2C). How the area is injured to cause cancer is not known. A squamous cell carcinoma may invade the connective tissue as a mass or in strands in which some groups undergo keratinization; others not differentiating are capable of doing so.

Basal cell carcinoma cells resemble those of the basal cell layer of the epidermis. The incidence ratio of basal to squamous cell carcinomas in light-skinned populations is about four in the northern United States, and somewhat less in the southern states. The incidence doubles with

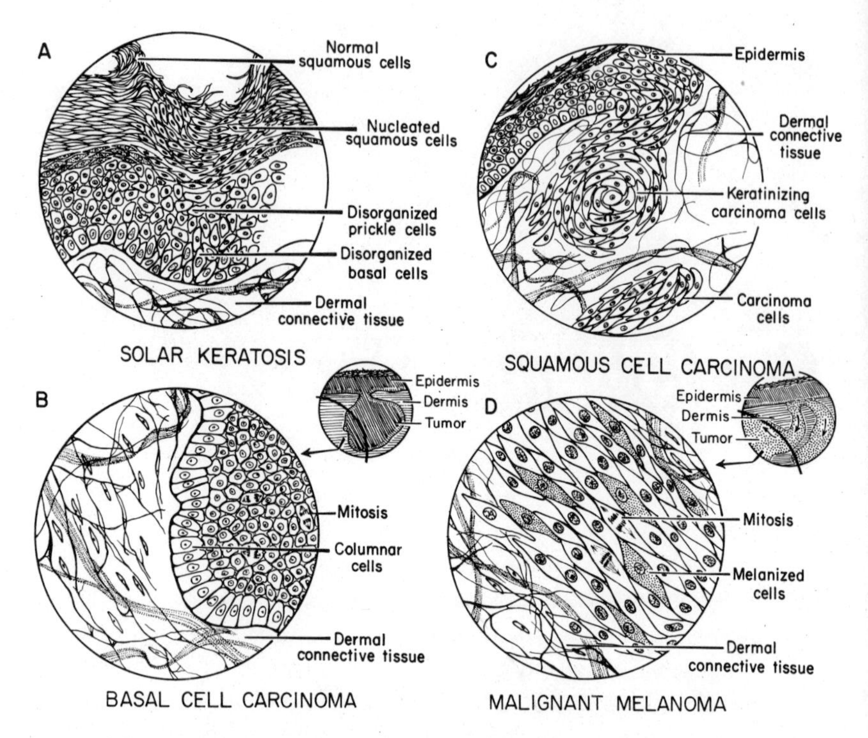

FIGURE 8:1. Illustrative diagrams of the tumors dealt with in this chapter. Partly after photographs in Milne, *An Introduction of Diagnostic Histopathology,* Williams and Wilkins, Baltimore 1972; partly from slides furnished by Dr. Steven Tomson, Dermatology Department, Stanford Medical School. (A) In solar keratosis more layers of squamous epithelium develop than in normal skin, with columns of fully keratinized cells alternating with columns of nucleated cells, under which there is considerable difference in cellular arrangement and structure. (B) Basal cell carcinoma as a nodule protruding into the dermis (small circle), part of which is enlarged to provide cell detail (large circle). The body of the tumor is delimited by a basement membrane and a layer of columnar cells. Cells in the body of the tumor may be of irregular sizes and some have abnormal nuclei, but mitoses are frequent. Some tumors of this type become cysts with cellular debris in the cavity, while others protrude as nodules on the skin surface. Basal cell carcinomas may ultimately ulcerate on the surface of the skin. (C) Squamous cell carcinoma as seen in section. Details of the epidermis shown to the left. Note the connection of the invading strands to the surface. The two strands have invaded the dermis and the cells are among the connective tissue fibers. While many of the cells remain relatively undifferentiated, some become keratinized, as seen in the nest at the top forming a "pearl" of keratinized cells. This is only one of many forms of the tumor. (D) Malignant melanoma invading the dermis (small circle) and details of some of the invading cells (large circle). While in early stages of the tumor the basal layer may be replaced by multiplying melanocytes, when these break through the basement membrane and invade the dermis, the cells become epithelial, losing their characteristic dendrites, but they retain the capacity to melanize. Not all the cells form melanin but enough of them do to give the tumor a dark color. Since invasion is diffuse, multiple tumors may result. (Melanocytes may also give rise to colorless or amelanotic tumors [see text].)

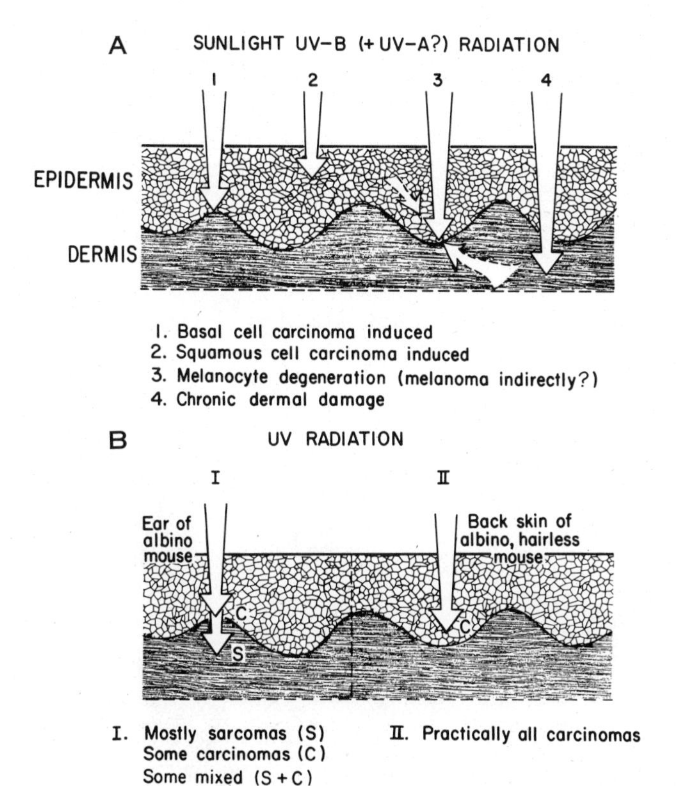

FIGURE 8:2. (A) Diagram of presumed effects of chronic overexposure of human skin at various levels: (1) basal cells, (2) squamous cells, (3) melanocytes, (4) dermis. (B) diagram of different carcinogenic effects of ultraviolet radiation on the skin from a hairy mouse (I) and a hairless mouse (II). The thinner epidermis of the hairy mouse ear permits deeper penetration than in the back skin of the hairless mouse, resulting in damage to the dermis. This induces sarcomas.

each 8 to 11° decrease in latitude. In the normal epidermal cycle, a division product of the basal cell spends a period of time in each layer of the epidermis as it matures and then dies to become part of the squamous cell layer (see Figure 5:2). In a basal cell carcinoma, on the other hand, the cells appear to be incapable of maturation and ultimately pile up to form a tumor. Biologists do not yet know what deficiency prevents cellular maturity. Interestingly, however, when dermatologists transplant a piece of a basal cell carcinoma to skin of the same person on a part of the body protected from the sun, the cells synthesize keratin and differentiate into normal-looking squamous cells. This suggests that sun-damaged cancerous cells possess the enzymes but lack the evocator for maturation normally supplied by the skin protected from sunlight. The basal cell carcinoma is bounded by columnar cells, longer cells bordering on the basement membrane. The other cells remain relatively undifferentiated. The tumor may form a solid mass of cells, or become a cyst with cells in the process of fragmentation and dissolution in the center.

Solar keratoses, often considered together with cancers to which they may be precursors, appear to the naked eye as hard papular growths or spreading flat growths with well-marked borders. At the cellular level they look like overdeveloped outer epidermis and consist of columns of normal squamous cells alternating with columns of less completely differentiated, flattened squamous cells retaining their nuclei. Below the latter columns the granular layer is missing, the prickle cells are disorganized, and the basal cells, which have lost their polarity, are increased in size with enlarged nuclei. The silvery-gray appearance of keratoses comes from the light reflected from air trapped between layers of squamous cells.

Dermatologists believe that malignant melanomas, usually dark-pigmented tumors of the skin that can take many forms, arise by multiplication of melanin-filled cells originating from melanocytes. The incidence of melanoma is about one-tenth that of squamous and basal cell carcinomas combined. Figure 8:1D depicts a melanoma. Like most malignant cancers, melanomas quickly spread to other sites, where they start secondary tumors, as shown in Figure 8:10, soon causing death. Human malignant melanomas may originate on skin in all areas of the body. They are not limited to, but most common on, areas most exposed to the sun. Dermatologists suggest that they are either induced directly by sunburn, or by chemicals diffusing from sunburned skin cells. The strongest evidence for a photochemical origin of skin melanomas is that in victims of xeroderma pigmentosum malignant melanomas are generally found on areas of the skin most exposed to sunlight, and they show an extraordinarily high prevalence (in one clinical sampling 7 to 15 percent by age ten) in such individuals compared to that in the general population (one in five thousand). The malignancy of melanoma is apparently not correlated with the presence of melanin in the cells, because melanocytes lacking enzymes to form melanin form a colorless (amelanotic) melanoma equal in invasiveness and spread, to a colored (melanotic) melanoma.

Animal Experiments

For the better part of the last century, scientists have consistently accumulated facts to show that the common denominators of cells are far more numerous than the differences between them. The basic biochemical cellular mechanism is so alike in the genetic code, DNA, RNA, and protein synthesis that we can use the several parts of a synthetic mechanism from different species even as distantly related as plants and animals, and carry out DNA synthesis in a test tube, for example, directed by the particular DNA template supplied from a human cell. Biologists therefore safely assume that the changes in cells of mice and other experimental animals involved in cancer induction by UV-B radiation are similar to changes in cells in human carcinogenesis. We may also infer that the changes in the skin as a milieu for the irradiated cells are probably similar in both kinds of skin. If these deductions are correct, then the data gathered on mice as well as other experimental animals are relevant to humans. In an analysis of basic mechanisms in radiation carcinogenesis, details may differ, because human skin cells and mouse skin cells may have different susceptibilities to ultraviolet radiation. Also there may be other undisclosed but concomitant variables peculiar to human carcinogenesis that are not revealed by experimental animals. (For example, compare Figure 8:2A with Figure 8:2B, I and II.)

In the laboratory, cancers can be induced by sunlight in the skin of albino rats. When researchers place window glass between the sun and the rat, the radiation does not induce growths, suggesting that only the UV-B rays are carcinogenetic. However, very large doses of UV-A radiation have recently been shown to be carcinogenetic. Since sunlight is neither available at all times nor controllable even when available, researchers use artificial sources of ultraviolet radiation. These include carbon and mercury arcs and, at present, xenon arcs. Some experimenters use the whole emission spectrum of the radiation source, while other researchers use only part of the spectrum transmitted by selected filters. Currently, some workers use monochromatic radiation. Experiments also show that high temperature during irradiation (not necessarily before and afterward) accelerates skin cancer.

Most skin cancer laboratories perform experiments with mice, rats, and guinea pigs. In most cases, albino animals are preferred because unpigmented skin is less resistant to ultraviolet radiation. Researchers have found that ultraviolet radiation readily induces tumors on the relatively hairless skin of the ears of albino mice. The quantitative relation between the dose of ultraviolet radiation and induction of skin tumors has been determined in this manner. Development of sarcomas (dermal tumors), as compared to carcinomas (epidermal tumors), following ultraviolet irradiation of the skin of a mouse ear implies that enough ultraviolet radiation was able to reach the susceptible cells in the dermis. Measurements show that the thin epidermis of the ear skin of the albino mouse transmits carcinogenetic UV-B radiation down to the dermis and tumors are therefore mostly sarcomas (see Figure 8:2B, I).

At present, however, experimenters prefer new strains of hairless albino mice and rats with thicker epidermis. In these animals the tumors are carcinomas induced by changes in the epidermal cells, much like the human skin tumors (see Figure 8:2B, II). In addition, the tumors can be induced on the backs of mice, where a known and repeatable dose of radiation is more easily administered than on ears. Hairless mice and rats are also used to show the progressive histological changes in irradiated skin cells, especially in the epidermis. These new strains have also been used to show changes in macromolecular syntheses after irradiation, the rate of mitosis in skin cells with incipient carcinogenesis after periodic irradiation, and so on. In experiments on melanoma induction, pigmented hairless mice have been used instead of albinos because pigmented cells are more easily observed.

Let us now consider the laws of photochemistry in relation to skin cancer. If skin cancer has a photochemical basis, ultraviolet carcinogenesis should be consistent with these photochemical rules. One states that photochemical reactions are only promoted by the light absorbed. Albino mouse skin absorbs UV-B radiation, but not much UV-A radiation or visible light; UV-B radiation was shown to induce cancer much more readily than UV-A radiation.

Another law states that, for a given total dose of radiation, the photochemical effect should be the same, regardless of whether the radiation is applied at a low intensity for a long time or a high intensity for a short time, a concept well known to photographers. This so-called reciprocity law has been shown to work in extensive experiments on ultraviolet radiation-induced tumors on the ears of albino mice. Thus, tumor incidence and prevalence in mouse ear skin was found to be the same for the same doses delivered either as low- or at high-intensity ultraviolet radiation, provided the interval between successive exposures and other

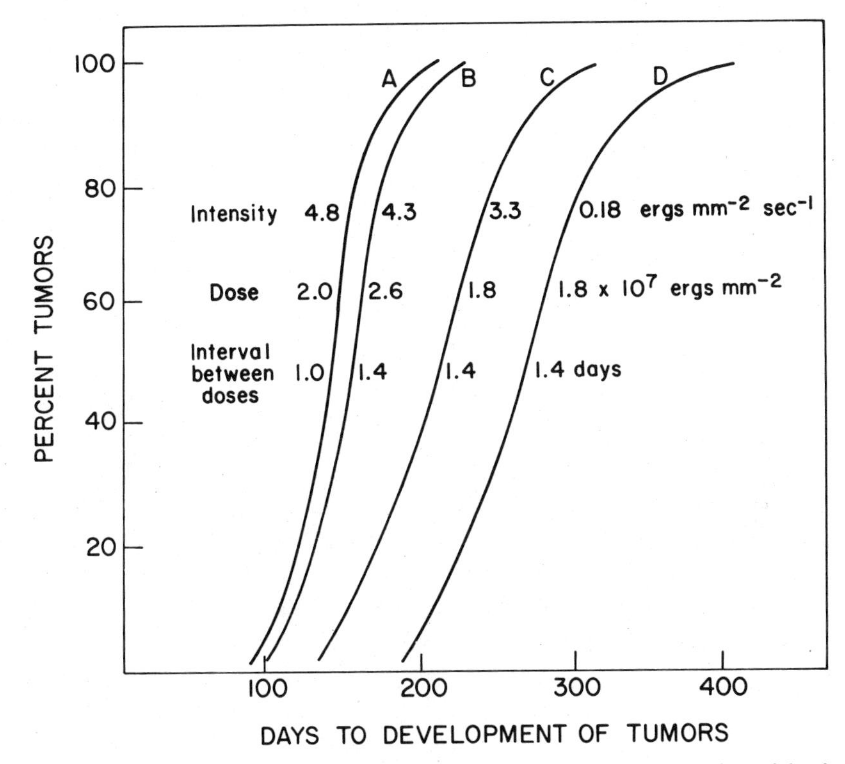

FIGURE 8:3. The prevalence of tumors on the ears of albino, hairy mice and time of development of the tumor with the following variables: difference in dose and interval between doses, but at the same intensity (curves A and B); difference in dose and intensity but at same time interval (curves B and C); difference in intensity but same dose and interval between exposures (curves C and D). After Blum, *Carcinogenesis by Ultraviolet Light,* Princeton University Press, 1959, p. 190.

conditions were the same. When the intensity was reduced to a very low value, however, carcinogenesis was delayed, even though the total dose was the same. Biologists believe that reciprocity failure results from active repair of ultraviolet radiation damage in the irradiated cells (see Chapter 4). Low intensity apparently provides more time for repair, as shown in Figure 8:3, curve D.

When researchers strike the ears of mice with a single threshold skin-damaging exposure to ultraviolet radiation, tumors do not appear. Obviously, a large number of doses are required. In one study, a minimum of fifty such exposures over seventy-four days produced tumors (Figure 8:4D). The rate of cancer induction is dependent on the dose of radiation; the larger the dose within limits, given in the same interval of

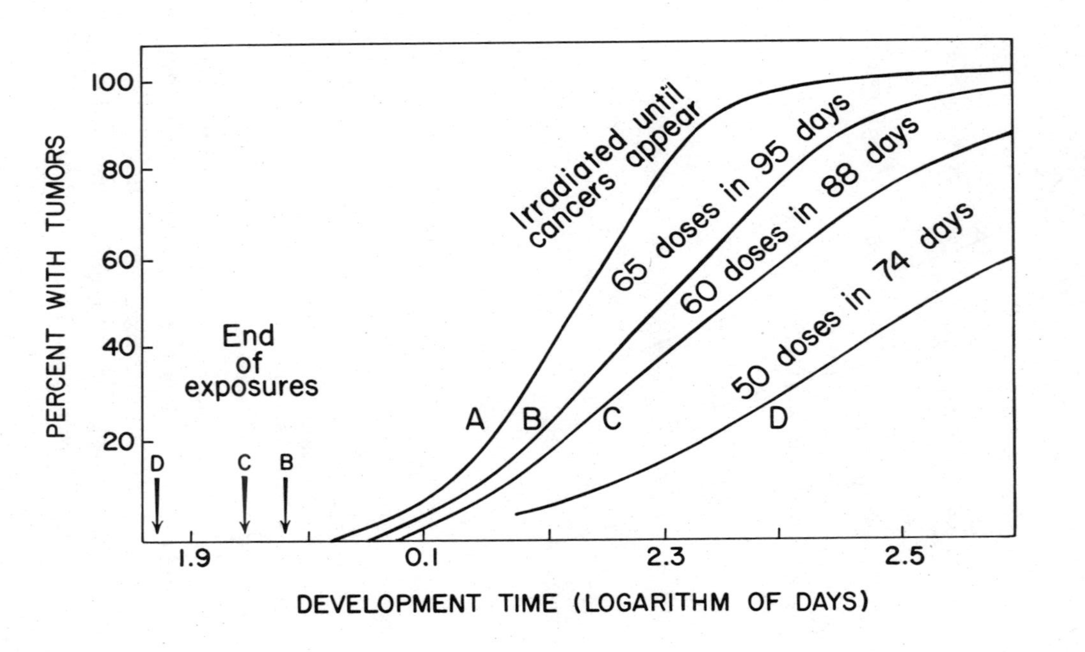

FIGURE 8:4. Delayed appearance of ear tumors on hairy albino mice as a result of discontinuance of irradiation at various times. The dosage was 1.8×10^7 ergs cm^{-2}, the interval between was 1.4 days (irradiation daily except Saturday and Sunday). The curve labeled A is for the population irradiated regularly until tumors had appeared. For the other curves, exposure was stopped at various times as indicated on the abscissa. To separate the curves better, the time for development of tumors is given as the logarithm of the time in days. After Blum, *Carcinogenesis by Ultraviolet Light,* Princeton University Press, 1959, p. 220.

time, the quicker the appearance of skin cancer (see Figure 8:4, curves B, C, and D). The rate of cancer induction is also influenced by the frequency of exposure to the same dose of ultraviolet radiation—daily, weekly, etc. The larger the time interval between successive exposures, the longer the time required for cancer induction (see Figure 8:3, curves A and B). Apparently, a finite total dose of ultraviolet radiation is required for carcinogenesis, but other factors may modify the time of tumor onset.

Although a single ultraviolet radiation exposure equal to one dose in a series of fifty, cumulatively carcinogenic, does not induce a tumor in mouse skin, it has been postulated that even a single dose starts skin cells on their carcinogenic road. Usually, periodic irradiation is continued until a cancer appears on the skin. But, if exposure to ultraviolet radiation is terminated before the tumor appears, carcinogenesis is only delayed (see Figure 8:4), with the delay being longer the sooner the treatment is terminated until the delay stretches close to the end of a mouse's lifetime. Theoretically, if a mouse lived long enough, perhaps it would have developed a tumor from a single dose of ultraviolet radiation. One experiment suggests that a single dose of ultraviolet radiation on a mouse ear initiates a tumor when the irradiated area is periodically treated with croton oil (a cocarcinogen extracted from the seed of a plant, *Cascarilla*), which by itself is not a carcinogen. Neither of the controls, one similarly treated only with croton oil, another with the same dose of ultraviolet radiation only, developed a tumor. It is assumed that the croton oil in some manner "develops" the change initiated by a single dose of ultraviolet radiation to the skin cells, thus substituting for the accelerating action of successive doses of ultraviolet radiation normally used for tumor induction. Scientists have yet to explain the mechanism by which the croton oil develops a tumor after such a small dose of radiation; nor do they know the mechanism by which successive treatment with ultraviolet radiation accelerates carcinogenesis initiated by the radiation. Croton oil, however, amplifies the effect of chemical carcinogens in a similar, undetermined manner.

In another experiment, researchers administered the total cumulative dose of ultraviolet radiation required to induce a skin tumor in fifty weeks in one massive dose to hairless albino mice. The damage to mouse skin cells was so great that about half of the mice died. Large ulcers appeared on the skin of the survivors. At the edges of the healing ulcers, experimenters could see numerous small basal cell carcinomas. That the single massive UV dose acts differently from the cumulative effect of a series of periodic small UV doses is shown by a difference in the final effect. Tumors induced by the series of small doses are not known to regress, whereas those induced by a single massive dose do. It is possible that the single massive dose so alters the milieu for skin cells as to induce a tumor chemically; some chemically-induced tumors are known to regress.

The molecular mechanism by which skin cancer in mice is photochemically induced by ultraviolet radiation remains obscure. Researchers have not yet unraveled the action spectrum that might provide the clue as to the target molecule for ultraviolet radiation carcinogenesis. Several obstacles stand in the way: Tumor induction requires several months of periodic treatments with relatively large doses of radiation, and the monochromatic radiation available for action spectrum studies is generally not only of low intensity but also covers a small area. Also, consider how difficult it is to confine a wiggly mouse to make sure that the monochromatic light is focused on the same area of skin each time. Available evidence from exposure of mice to filtered polychromatic radiation and preliminary experiments with monochromatic radiation suggests that the UV-B radiation band (286–320 nm) is most effective in inducing skin cancers similar to those induced by sunlight. Scant evidence suggests that the carcinogenic action spectrum is similar to that for sunburn. Investigators interpret this as an effect on cellular nucleic acid or acids. They believe that DNA is the target molecule because of the changes in its replication in radiation-damaged cells (first an inhibition, then an acceleration) after ultraviolet irradiation. UV-C wavelengths are also carcinogenic.

Biologists are far from deciding how ultraviolet radiation induces cancer. However, they have narrowed the choices down to four: (1) somatic mutations, (2) release of irradiated cells from division-inhibiting compounds (so-called chalones and possibly other regulators) synthesized in the skin, (3) alteration of immune mechanisms, or (4) permitting multiplication of viruses that induce cancer in injured skin cells. All of these theories are also proposed explanations for how, in general, cancer develops.

Ultraviolet radiation induces mutants in many types of cells and DNA is the target molecule (see Chapter 4). However, no one yet knows whether the ultraviolet-induced carcinogenesis is the result of ultraviolet-induced mutagenesis in mammals.

Recent studies indicate that a cancerous change can be induced in whitefish cells from exposure to ultraviolet rays. Ultraviolet radiation alters DNA, predominantly by inducing thymine dimers in the DNA. The dimers are split if the cells are subsequently exposed to visible and UV-A radiation (photoreactivation), and most of the DNA then returns to its native state. If radiation-induced carcinogenesis depends on dimer formation in DNA, it should be subject to photoreactivation. Such was indeed found to be the case. This unique experiment performed by researchers at the Oak Ridge National Laboratory represents the single most convincing evidence that UV-radiation induces cancer in animal cells.

Sunburn Injury and Human Carcinogenesis

Since no experimental data exist on human carcinogenesis by ultraviolet radiation, all evidence of whether there is a cause-and-effect relation between sunburn damage and human skin-tumor generation remains mainly circumstantial. It depends on statistical correlations between incidence and prevalence of cancer and sun exposure. Unfortunately, not all clinical data for human skin cancer have been properly collected. In the past, quite a few physicians did not check their diagnoses of skin cancer by histological examination of tumor cells. Moreover, patients themselves had to estimate their cumulative exposure to sunlight, a difficult task for any of us. Also, from physicians' clinical records, statisticians cannot easily assess melanin content of a patient's skin and ease of tanning. To add to the difficulty, doctors report skin cancers in a variety of ways: Some note each tumor that appears during a time span (usually six months or a year) as a separate case; others list each cancer patient as an individual case, regardless of the number of tumors each one has. Despite these obstacles, cancer induction in human skin nevertheless appears to be correlated with sunlight exposure.

If biologists are correct in assuming that ultraviolet radiation is carcinogenic, there should be a correlation between exposure to sunlight and the incidence and prevalence of human skin cancer. Most cancers would then be found in (1) parts of the body most exposed to sunlight, (2) those who spend the most time outdoors in the sun, (3) people with similar heredity at lower latitude and in higher altitudes, (4) other factors being the same, older people more than young, because of their longer cumulative period of exposure to the sun, (5) light-skinned individuals more than dark-skinned people, and (6) those with light skins who do not tan more than those who do.

What Parts of the Body Are Prone to Skin Cancer? As mentioned before, clinical surveys reveal that human skin cancers appear on areas of the body receiving greatest exposure to sunlight, namely the head and neck. In one representative survey, 840 basal-cell carcinomas, all verified by histological examination, were found to be about equal on both sides of the body; 421 on the right side, 419 on the left. Of these 91.2 percent were located on the head and neck, and less than 9 percent occurred on the rest of the body. Figure 8:5 illustrates that most of the tumors were found on the nose and adjacent portions of the face. Several times as

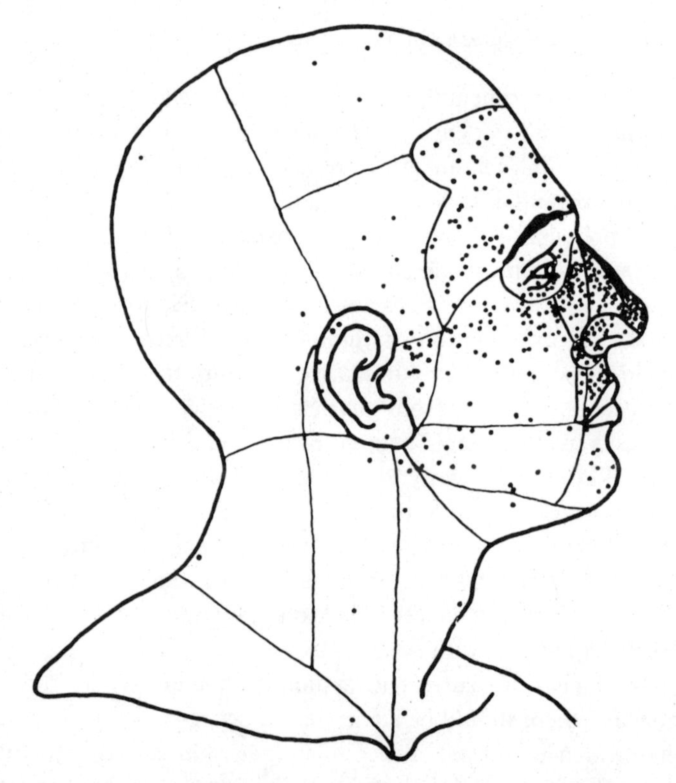

FIGURE 8:5. Localization of skin cancers on the human head, ears, and neck, the relative number being indicated by the number of dots. From Brodkin, Kopf, and Andrade, in *Biologic Effects of Ultraviolet Radiation,* edited by F. Urbach, Pergamon Press, Oxford, 1969, p. 586.

many carcinomas occurred on the lower as on the upper eyelids. A surprisingly greater number of tumors, twelve to one, was present on the ears of men than women. It must be pointed out, however, that the survey was done at a time when women wore their hair longer than men. On areas other than the face, more tumors occurred on both front and back surfaces of the chest than on the belly and rump, and more on the lower legs of women than men. When dermatologists surveyed larger numbers of skin cancer victims, they found essentially the same results. Some skin specialists believe that some skin areas—for example, the bald scalps of men and the upper surface of the appendages in both sexes—resist cancer more than other parts of the body. Be that as it may, the mass of clinical evidence shows that the more the skin is exposed to the sun, the more there is a chance of skin cancer.

Work and Leisure. Skin cancer specialists have long observed that people who work outdoors are more likely to fall victims to skin cancer than those who do not. Studies show that farmers, sailors, teamsters, outdoor construction workers, and policemen have a higher incidence of skin cancer than office workers, clerks in stores, indoor manufacturers, and houseworkers. With the mechanization of farming and industry, many who use covered power machinery are now protected from the sun and in the future are much less likely to develop skin cancer.

As work hours shorten, there is more time for leisure activity, much of it outdoors and in milder climates, often with little clothing protecting the skin. In Australia, outdoor activity during a great part of the year is correlated with increased incidence of skin cancer.

Latitude. People at lower, rather than higher, latitudes are more likely to receive a greater cumulative exposure to sunlight. More hours of sunlight and warm climate encourage outdoor work and leisure activity. One survey compares skin cancer found in racially matched populations in low and high latitudes. As Figure 8:6 shows, it found a greater cancer incidence at low latitudes than at high ones. Other surveys agree with this general conclusion, though the findings are not identical quantitatively. More searching and systematic analysis is needed to determine the cause of the quantitative differences. Interestingly, prevalence of malignant melanomas is distinctly correlated with latitude, even though melanomas occur even on parts of the body not generally exposed to the sun.

Age. After people reach age fifty, squamous cell carcinomas and basal cell carcinomas, lumped together, and the more common keratoses (see Figure 8:7) show a sharp rise. Dermatologists believe that this

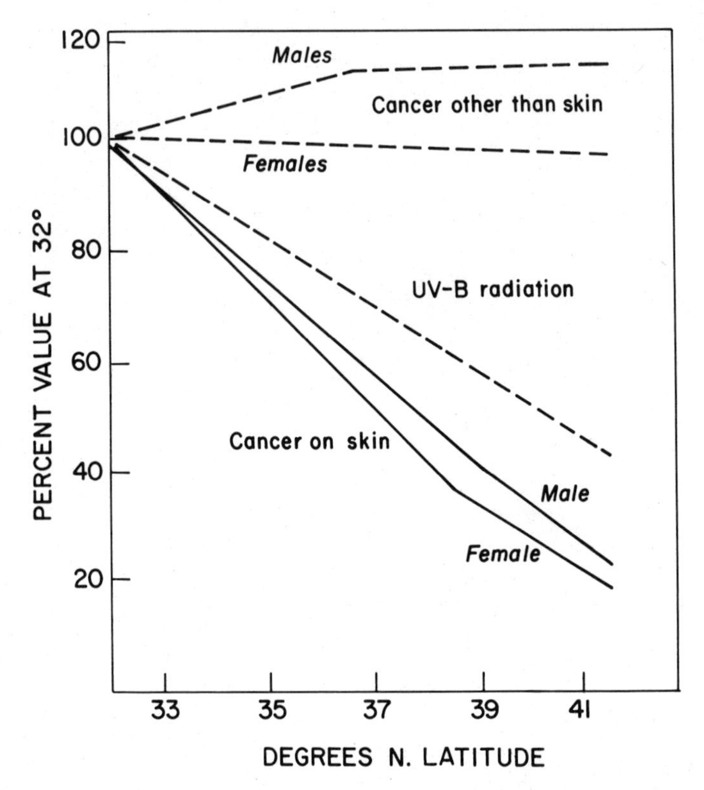

FIGURE 8:6. Variation with latitude of carcinogenic UV-B radiation of sunlight (middle curve) and incidence of skin cancer (lower two curves). The estimated maximum variation in carcinogenic radiation of sunlight was not corrected for scattering that would diminish the variation with latitude. The correlation in cancer incidence with total exposure to UV-B radiation as a function of latitude is as significant as is the lack of any such correlation in other types of cancer (top two curves) with variation in latitude. After Blum, *Carcinogenesis by Ultraviolet Light,* Princeton University Press, 1959, p. 296.

happens perhaps as a consequence of age alone, or as a consequence of cumulative exposure to the sun during one's lifetime, or possibly as a combination of both. Some workers conjecture that skin cancer in older people may be caused by failure of the regulatory mechanisms preventing cell division of cells other than germinative ones. Uncontrolled cell division would lead to a disorganized mass of cells characterizing cancer.

If skin cancer is a result of aging alone, with its assumed loss of cell division control, then it should appear the same in populations of the same genetic types at high as well as low latitudes. If, on the other hand,

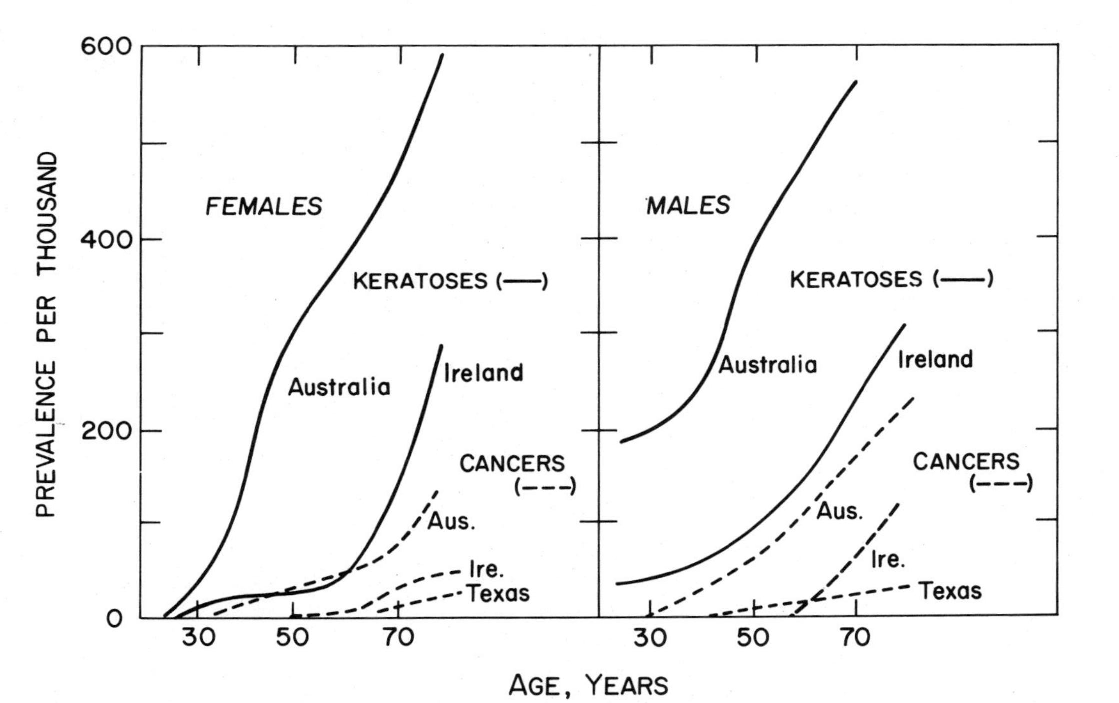

FIGURE 8:7. Increased prevalence of skin cancer and keratoses with age of the populations from different latitudes and of different genotype. Caboolture, Australia, is at 27.5° south latitude, Galway, Ireland, is at 53.3° north latitude, and both have a predominantly Celtic population; El Paso, Texas, is at 31.8° north latitude, but the population is predominantly non-Celtic. It is also evident that both kerotoses and skin cancers (data for the sum of both basal cell carcinomas and squamous cell carcinomas) appear earlier in the predominantly Celtic population, and incidence is higher. After Urbach, Rose, and Bonnem, "Genetic and Environmental Interactions in Skin Cancer," in *Environment and Cancer,* Williams and Wilkins, Baltimore, Md., 1972, p. 367.

cancer is a result of damaging exposure to the sun, then it should increase with greater cumulative exposure. The data indicate that skin cancer is most prevalent at low latitudes with the greatest exposure to sunlight (see Figure 8:7) regardless of age. While age is unquestionably a factor in skin carcinogenesis, it appears that both incidence and prevalence of skin cancer increase with cumulative exposure to sunlight.

Cumulative Exposure. If records existed on the UV-B radiation on each day of the year in areas with comparable populations, at high, intermediate, and low latitudes, researchers could calculate the number of minimal erythemal doses on each day. They could then correlate the incidence and prevalence of skin cancer and the cumulative UV-B dose in minimal erythemal units for the skin cancer surveys already made. Dermatologists could then correlate skin cancer and UV-B radiation, knowing the total cumulative UV-B radiation in more meaningful units; it would then be pointless to include doses on days with less than one minimal erythemal dose. At some latitudes—for example, Minneapolis at certain seasons—there are many days when the sunlight may fall below even a single minimal erythemal dose. On these days, recovery from the slight effects of the skin is probably almost complete. Today, continuous photocell meters measure the relative erythemal effectiveness of sunlight from UV-B rays at stations in key locations in the United States and abroad. After these measurements are completed, it will be possible to correlate the data with skin cancer incidence and prevalence and we should have more definitive answers than now (see Note Added in Proof, p. 150).

Race and Genetics. We have mentioned that physicians noted long ago that skin cancer is much less frequent in dark-skinned people than in those with light skins. Statistics show that black-, brown-, and yellow-skinned people are much less susceptible to cancer than Caucasians. Olive-skinned Caucasians, in turn, are less susceptible than those with lighter skin. Light-skinned people who tan are less susceptible than those who do not tan or only freckle, as, for example, the Celts. Australia, with a large proportion of people of Celtic origin, shows a higher prevalence of skin cancer than other countries with predominantly light-skinned populations but with a lower percentage of people of Celtic descent. Supposedly, albinos are even more prone to skin carcinogenesis. For example, albino Cuña Indians of Panama are much more prone to skin cancer than their dark-skinned siblings, but the samples are small

FIGURE 8:8. Albino "moon-children" of the Cuña Indians of Panama stand in contrast to their dark-skinned mother. Though a recessive, the albino type appears frequently in the population because of inbreeding. Courtesy of Dr. Clyde Keeler, Director, Research Division, Central State Hospital, Midgeville, Georgia.

(Figure 8:8). Albinos are less frequent in other racial groups than in this highly inbred population and consequently the data for them are even more meager. Obviously, skin pigment is of great value for protection against skin carcinogenesis.

That dark repair is perhaps of even greater importance than pigment in protecting against skin carcinogenesis is shown by the extraordinary susceptibility of victims of xeroderma pigmentosum, who lack either the excision repair or the postreplication repair system (see Figures 4:13B and 8:9). Regardless of skin pigment, many with this disorder have

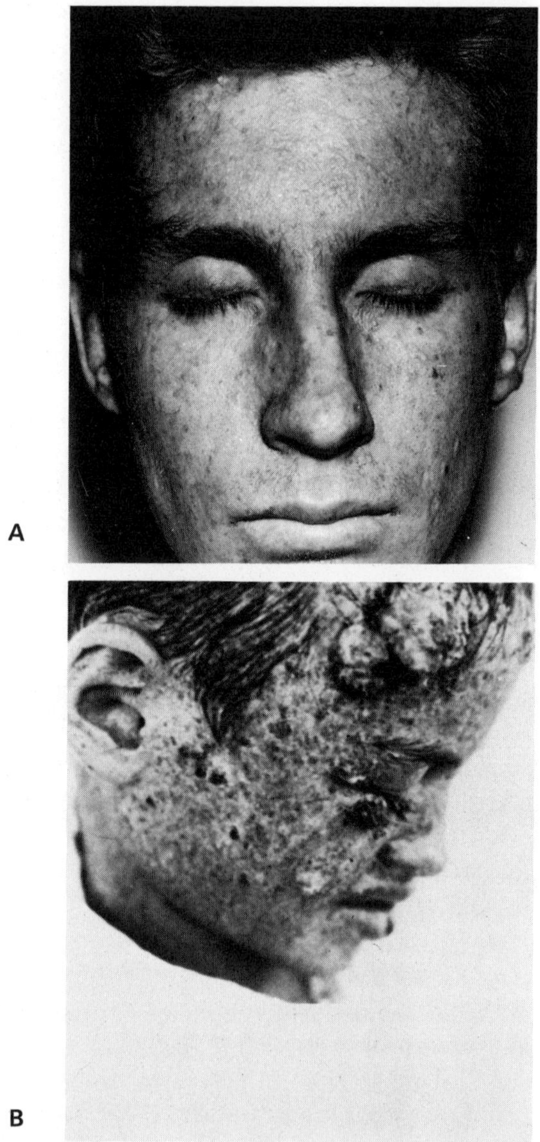

FIGURE 8:9. (A) An early case of xeroderma pigmentosum. Note the spotty pigmentation and the lesions on the cheek. Courtesy of Dr. Frederick Urbach, Chairman, Skin and Cancer Hospital, Temple University Medical School, Philadelphia, Pa. (B) A more advanced case of xeroderma pigmentosum showing numerous keratoses and a large squamous cell carcinoma. Courtesy of William Spencer, M.D., from Epstein, *Sunlight and Man,* edited by T. Fitzpatrick *et al.,* Tokyo University Press, Tokyo, 1974, p. 300.

skin cancer at age two and most are condemned to die of cancer in their teens. This only emphasizes again the precarious balance by which we survive in an environment in which UV-B radiation from the sun is continuously damaging the skin.

Relative Malignancy of Skin Cancers

During early animal development, cells multiply and wander actively for a time. Later, on making contact with one another, they adhere (contact inhibition), stop dividing, and differentiate. Tumor cells, on the other hand, glide over one another (in tissue culture and presumably in animals). They continue to divide but do not differentiate. Scientists are still mystified by the nature of the molecular changes that lead the cells to behave as tumor cells. Multiplication continues until secondary factors not yet fully defined intervene. By this time, a large mass of undifferentiated and unorganized cells forms. Since effective circulation does not always develop, cells lacking adequate oxygen resort to anaerobic metabolism (glycolysis). Older established tumors usually contain dying cells in their interior.

Even though tumors may remain confined, nonetheless they may enlarge, causing incidental damage by pressure on nearby tissue or vessels and ducts, sometimes even causing death. Some tumors spread and invade other tissues with increasingly damaging effects, uncontrolled by body reactions. When a primary tumor spreads (metastasizes) away from its original site to other locations in the body through the blood or lymphatic systems, colonies of cells soon form secondary tumors elsewhere (Figure 8:10). When this happens, surgical or x-ray procedures become ineffective for control. Obviously, benign skin tumors, sharply defined from surrounding tissues, do not transfer to another host; malignant tumors, on the other hand, with no distinct boundaries, do transfer.

Normally, keratoses are confined, as are many basal cell carcinomas. Squamous cell carcinomas usually invade the dermis and may later metastasize. Melanomas spread quickly and metastasize, damaging other tissues (Figure 8:10). It is best to remove skin tumors as soon as they are discerned because ultimately even keratoses may develop into squamous cell carcinomas.

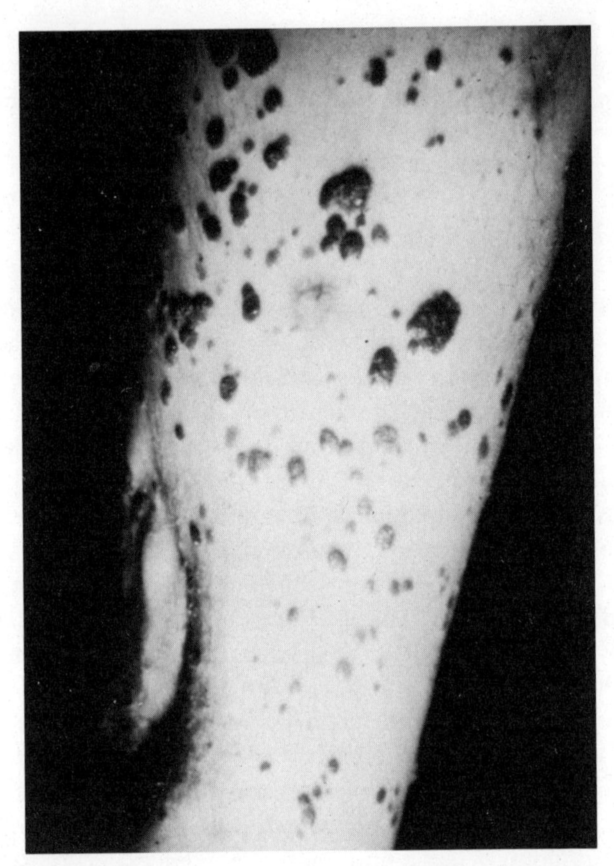

FIGURE 8:10. Secondary melanomas on the arm of an individual developing as a result of the metastasis of a primary malignant melanoma. By courtesy of the Dermatology Department, Stanford University Medical School, Palo Alto, California.

FIGURE 8:11. Structure of the carcinogenic compound 3,4-benzpyrene, a fluorescent hydrocarbon found in coal tar, a potential carcinogen for man. Its high fluorescence indicates uptake of ultraviolet radiation and emission at visible wavelengths. Not all fluorescent compounds are carcinogenic.

Photosensitized Reactions to UV-A Rays

Physicians have known for years that chronic exposure to coal tar, pitch, and similar substances induces skin cancer. Because its fluorescence was so strong, 3,4-benzpyrene, a polycyclic hydrocarbon (as indicated in Figure 8:11), was early isolated from coal tar. It has been shown to be an important carcinogen. Formed in the high-temperature decomposition (pyrolysis) of carbonaceous compounds, it is a product of incomplete combustion—in diesel exhaust and cigarette smoke, for example. After chemists isolated this substance, other hydrocarbons present as products of pyrolysis were quickly discovered, many of them with carcinogenic action.

Strong fluorescence of some of these polycyclic hydrocarbons indicates that they absorb UV-A rays and re-emit some of the energy as visible light. When applied to the skin, these compounds absorb and some of them presumably transmit the light energy to skin cells. Researchers have tested this in some cases. For example, benzpyrine skin cancer is induced earlier in illuminated animals than in dark controls, but with some polycyclic compounds, light exposure has no potentiating effect. Pyrolytic decomposition of cholesterol, the common structural sterol in cells of vertebrates, leads to formation of another carcinogenic fluorescent hydrocarbon, methyl cholanthrene. A variety of other photodynamic compounds have been tested for cancer production in experimental animals.

Although the data are still sparse, most biologists believe that there always exists the possibility of potentiation of the carcinogenic action of sunlight by photosensitized damage in the presence of some of these substances on skin cells.

For Additional Reading

Blum, H. F. *Carcinogenesis by Ultraviolet Light,* Princeton, N.J.: Princeton University Press, 1959.

Cleaver, J. E. "Xeroderma Pigmentosum—Progress and Regress." (Review) *Journal of Investigative Dermatology* 60 (1973): 374–380.

Daive, C. J., and Harshberger, J. C. eds. *A Symposium on Neoplasms and Related Disorders of Invertebrate and Lower Vertebrate Animals.* Bethesda, Md.: National Cancer Institute Monograph 31, 1968.

*Epstein, J. H. "Ultraviolet Carcinogenesis." In *Photophysiology,* vol. 5, edited by A. C. Giese. New York: Academic Press, 1970, pp. 235–273.

Fitzpatrick, T. B., Pathak, M. A., Harber, L. C., Seiji, M. and Kukita, A., eds. *Sunlight and Man,* Tokyo: Tokyo University Press, pp. 259–283, 285–298, 299–315.
Freeman, R. G. "Recurrent Skin Cancers" *Archives of Dermatology* 107 (1973): 394–399.
Giese, A. C. "Ultraviolet Action Spectra in Perspective: with Special Reference to Mutation." *Photochemistry and Photobiology* 8 (1968): 527–546.
Hsu, J. Forbes, P. D., Harber, L. C., and Lakow, E. "Induction of Skin Tumors in Hairless Mice by a Single Exposure to UV Radiation." *Photochemistry and Photobiology,* 21 (1975): 185–188.
Keeler, C. "Albinism, Xeroderma Pigmentosum and Skin Cancer." National Cancer Institute Monograph 10 (1964): 349–359.
Lee, J. A. H., and Merrill, J. M. "Sunlight and the Etiology of Malignant Melanoma: A Synthesis." *Medical Journal of Australia* 2 (1970): 846–851.
Milne, J. A. *An Introduction to the Diagnostic Histopathology of the Skin.* Baltimore: Williams and Wilkins, 1972.
Robbins, J. H., Kraemer, K. H., Lutzner, M. A., Festoff, B. W., and Coon, H. G. "Xeroderma Pigmentosum. An Inherited Disease with Sun Sensitivity, Multiple Cutaneous Neoplasms and Abnormal DNA-repair."*Annals of Internal Medicine* 80(1974):221–248.
Schultz, J., and Graizer, H. G. eds. *The Role of Cyclic Nucleotides in Carcinogenesis.* New York: Academic Press, 1973.
Urbach, F. ed. *The Biologic Effects of Ultraviolet Radiation.* Oxford, Pergamon Press: 1969, pp. 433–435, 619–623, 635–650.
———, Rose, D. B., and Bonnem, M. "Genetic and Environmental Interactions in Skin Carcinogenesis." In *Environmental Cancer,* Baltimore, Md.: Williams and Wilkins, 1972.

Note Added in Proof

Too late for inclusion in the text or figures of this chapter a report prepared by J. Scotto, T. R. Fears, and G. B. Gori for the National Cancer Institute entitled "Measurements of Ultraviolet in the United States and Comparisons with Skin Cancer Data" which just appeared (1975) cites data (pages 3.1–3.10) clearly showing a relatively close correlation between the annual ultraviolet radiation dose in sunburn units and the rate of both non-melanotic and melanotic cancers.

9

How Atmospheric Pollution Affects Sunlight and Life

As seen from a satellite, the atmosphere wraps around the earth like a gaseous membrane, thin in comparison to the earth's diameter yet protective of life and constantly in flux because of natural processes and human activities. Since the Industrial Revolution we have dumped gaseous and fine particulate wastes into the air in even-increasing quantities without a thought. But, just like land and water, the atmosphere is finite. It can absorb only so much waste without losing its protective function.

Our atmosphere is densest at the surface of the earth and gradually thins above the surface until it merges into space (the exosphere). As depicted in Figure 9:1, based on temperature changes, the atmosphere can be subdivided into four layers: first, the troposphere close to the surface with a gradually falling temperature; next, the stratosphere just above it with a gradually rising temperature; still higher, the mesosphere with a gradually falling temperature; and finally, the thermosphere at the top with a gradually rising temperature.

Not surprisingly, atmospheric layers closest to us concern us most. Of particular note are the troposphere, which industrial smog affects seriously, and the second layer, the stratosphere, which might become seriously polluted by a projected commercial fleet of supersonic transport planes (SST), propellant gases from spray cans, nuclear explosions, and other human activity.

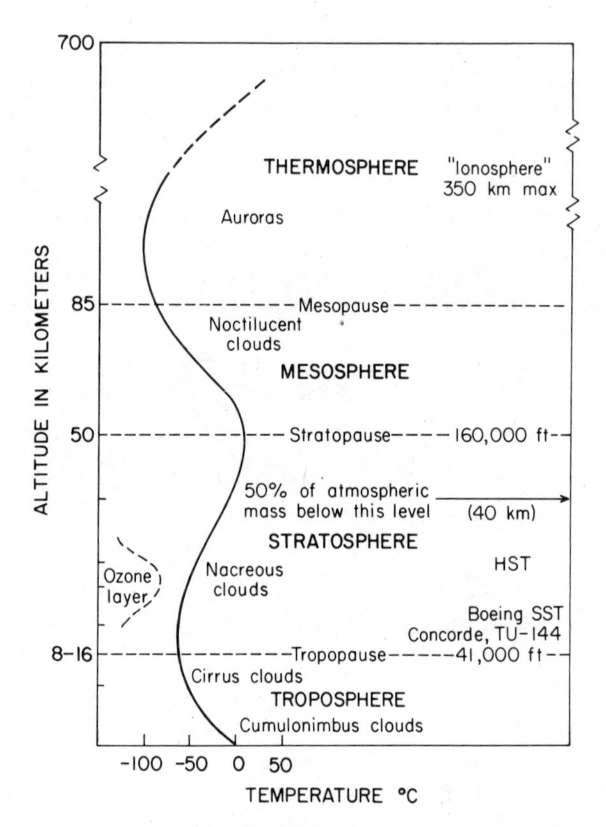

FIGURE 9:1. The stratification of the earth's atmosphere. The height of the tropopause varies with latitude (from 8 km at high latitude to 16 km at low latitude), seasons, and various atmospheric conditions. The vertical sinuous solid curve is the temperature curve. The ozone layer at midlatitude indicated begins about 15 kilometers above sea level and reaches a peak 25–30 kilometers above sea level. Abbreviations: SST, supersonic transport; HST, hypersonic transport; Tu-144, Soviet Tupolev SST. Atmospheric temperature at the earth's surface, shown arbitrarily as 0°C, varies with location and season.

The troposphere exchanges constituents with the earth's surface. The tropopause, its upper boundary, rises with decreases in the earth's latitude. At midlatitudes, it reaches about twelve kilometers above the earth. The troposphere, which receives its heat primarily by reflection and radiation from the earth's surface rather than directly from the sun (see Figure 9:2), varies in temperature and is a region of storms and turbulence. Temperature gradually drops with altitude, until at the tropopause it reaches a minimum above which it is stable, then rises again.

Because of the temperature inversion, with the cold gas mass below not rising and the warm gas mass above not sinking, little vertical mixing exists between the troposphere and the stratosphere above it, except at midlatitude tropopause gaps where jet-streams induce mixing between the two, and elsewhere only when violent electric storms near the boundary of the two layers inject air into the stratosphere. Exchange of

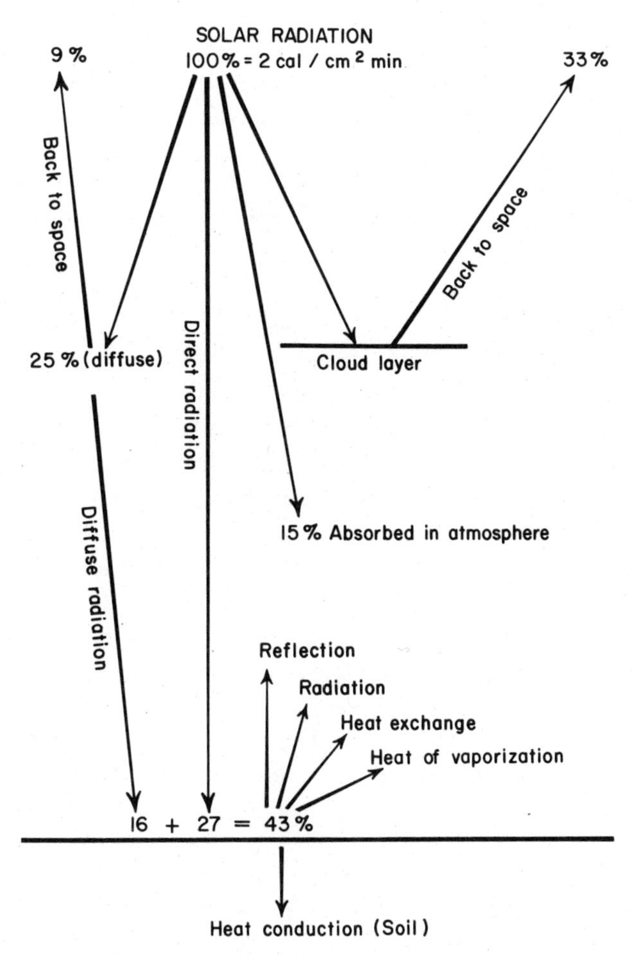

FIGURE 9:2. Heat exchange at the earth's surface and in the atmosphere at midday; the numbers represent average values. Diffuse radiation is that which is scattered from molecules and particles in the sky. From Allen, *Comparative Biochemistry,* vol. I, edited by Florkin and Mason, Academic Press, New York, 1960, p. 487.

gases in both directions between the two layers over most of the earth thus occurs primarily by diffusion.

The stratosphere extends from the tropopause to about fifty kilometers above the earth's surface. Approximately 99 percent of the air mass is below forty kilometers. Oxygen high in the stratosphere above the ozone layer absorbs the incoming UV-C radiation, forming ozone that is transferred toward the poles and downward, reaching a peak concentration at about twenty-five kilometers at midlatitudes (see Figures 9:1 and 9:7). Ozone, between fifteen and thirty-five kilometers above the earth, in turn absorbs some of the longer wavelength UV-C rays, and part of it is thereby decomposed back to molecular oxygen and atomic oxygen:

$$O_2 \xrightarrow[\text{peak 150 nm}]{\text{UV-C radiation}} 2O$$

$$O + O_2 \longrightarrow O_3$$

$$O_3 \xrightarrow[\text{peak 260 nm}]{\text{UV-C radiation}} O_2 + O$$

Much of the atomic oxygen immediately combines again with molecular oxygen to reform ozone, as in the second reaction. In the course of these reciprocal reactions, the energy of the absorbed ultraviolet radiation is dissipated as heat, gradually raising the stratospheric temperature to a maximum at the stratopause, the boundary between the stratosphere and the mesosphere above it.

Since the stratosphere has little exchange with the troposphere below, except by diffusion at the tropopause between the two layers, and diffusion is a slow process, any gas introduced into the stratosphere tends to remain in it for a year or more. Exchange is more rapid the closer the introduced substance is to the tropopause. Within the stratosphere, there is considerable east-west zonal circulation that spreads contaminants (debris from volcanoes, test substances, atomic bomb material, etc.) in a matter of days or weeks. The rather slow north-south circulation, dependent upon the meandering of the east-west zonal winds, spreads contaminants between northern and southern hemispheres in several months.

The mesosphere extends upward from the stratosphere to about eighty-five kilometers above the earth's surface. The temperature gradually falls to a second minimum as its upper boundary, the mesopause, is approached.

The thermosphere above the mesosphere extends out to about seven hundred kilometers above the earth, thinning at the upper reaches into space without a distinct boundary. The increase in temperature in this region by day is the result of absorption by air of the sun's short wavelength ultraviolet radiation and x-rays. This causes ionization of oxygen and nitrogen. By night, the higher temperature of the thermosphere is maintained primarily by energy from the charged particles of solar wind and the two van Allen belts of the earth, at about 3000 and 16,000 kilometers above the equator. Energy-rich particles, probably originating in the sun, are trapped by the earth's magnetic field.

Natural Tropospheric Smog

Smog (a contraction of smoke and fog), a fine suspension of material in air, generally consists of moisture particles (fog) containing dissolved chemicals and solid smoke particles. Smoke varies in color from blue, seen in natural haze and resulting from scattering of the short wavelengths of visible light by fine particulate matter, to yellow, brown, and black, resulting from reflection of only certain wavelengths of light, or none, from the suspended particles. Fog is white because water droplets reflect all the light striking them. Smog results from such natural processes as volcanic eruptions, forest fires, and operation of the cycles of matter.

Sulfur and sulfur compounds—hydrogen sulfide (H_2S) with the odor of rotten eggs, sulfur dioxide (SO_2), and sulfur trioxide (SO_3)—are given off by volcanoes and fumaroles. The same compounds are given off during the operation of the sulfur cycle, as shown in Figure 9:3. Sulfur dioxide and sulfur trioxide are also given off during forest fires. Hydrogen sulfide and sulfur dioxide are oxidized to sulfur trioxide by atomic and molecular oxygen and by ozone present in the atmosphere, but most of all by nitric oxide (NO) present in natural smog. Sulfur trioxide dissolved in water produces sulfuric acid (H_2SO_4), which, together with sulfur dioxide and sulfur trioxide, are acrid irritants to eyes and nasal passages. Sulfuric acid reacts with minerals to form sulfates (for example, sodium sulfate, Na_2SO_4) that appear in ocean spray.

Nitrogen compounds, nitrous oxide (N_2O), and ammonia (NH_3) are also present in the atmosphere, primarily as a result of the operation of the nitrogen cycle illustrated by Figure 9:4. Nitrous oxide is known to

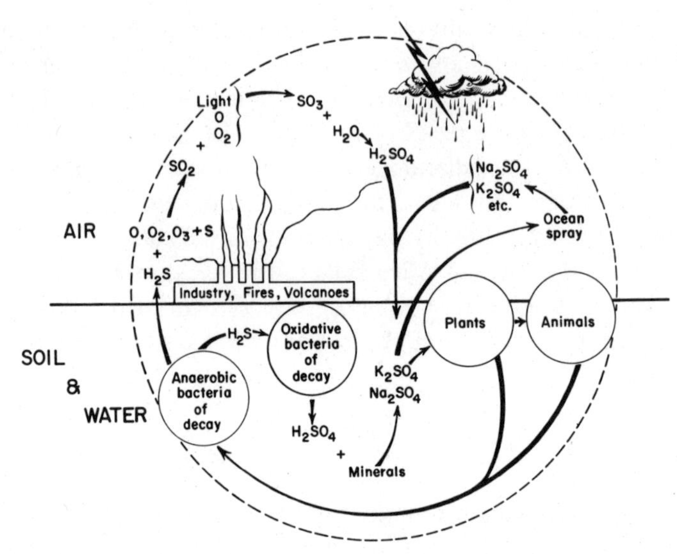

FIGURE 9:3. The sulfur cycle showing schematically the chemical processes involved in the transformation of sulfur as it moves from compound to compound in organisms and in natural waters, soils, and air. Legend: H_2S, hydrogen sulfide; O, atomic oxygen; O_2, molecular oxygen; O_3, ozone; SO_2, sulfur dioxide; SO_3, sulfur trioxide; H_2O, water; H_2SO_4, sulfuric acid; Na_2SO_4, sodium sulfate; K_2SO_4, potassium sulfate; sulfur (S) issuing from volcanoes (see Table 2:3) and from some industrial processes (in gaseous form as S_8 or S_2, depending on temperature) is oxidized to SO_2. Data from Kellogg *et al.*, *Science* 175 (1972), p. 588.

be released from the surface of the ocean and from rich wet soil by organisms of the nitrogen cycle, and is found at a concentration of 0.3 to 0.5 parts per million in the troposphere. Ammonia, released into air by organisms of the nitrogen cycle, reacts with the acidic compounds in the troposphere to form ammonium salts, and both these and ammonia dissolve in raindrops and are thus scrubbed from the atmosphere by rain. Otherwise, ammonia is stable in the troposphere. When organic nitrogen compounds present in trees are burned in forest fires, nitric oxide (NO) is released and rapidly oxidized to nitrogen dioxide (NO_2) by the ozone present in low concentration in tropospheric air. Nitrogen dioxide in turn is decomposed by absorption of ultraviolet radiation to form nitric oxide and atomic oxygen, and the atomic oxygen reacts rapidly with oxygen molecules to form ozone, thus completing the cycle. Steady states of ozone and NO may at times be maintained in the troposphere.

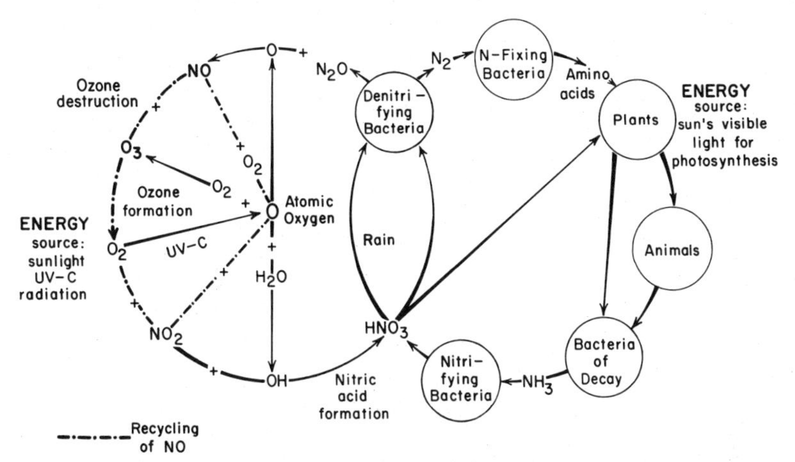

FIGURE 9:4. The nitrogen cycle (right) with the newly recognized secondary cycle (left) involving stratospheric nitrogen oxides (N_2O, nitrous oxide; NO, nitric oxide; NO_2, nitrogen dioxide; HNO_3, nitric acid), nitrogen (N_2), ammonia (NH_3), ozone (O_3), molecular oxygen (O_2), atomic oxygen (O), water (H_2O), and hydroxyl radical (OH). Data from Johnston, Proceedings of the National Academy of Science, U.S.A. 69 (1972), p. 2371.

Table 9:1. Naturally Occurring Compounds in Tropospheric Air[a]

Compound	Chemical formula	Compound	Chemical formula
Methane	(CH_4)	Hydrocarbons[b]	C_nH_{2n+2}
Butane	$(CH_3CH_2CH_2CH_3)$	Esters[c]	$CH_3COOC_2H_5$[d]
n-Butanol	$(CH_3CH_2CH_2CH_2OH)$	Aldehydes[e]	(RCHO)
Acetone	$(CH_3CO_2CH_3)$	Ketones[e]	(RR′CO)
Ammonia	NH_3	Ozone	O_3
Nitrous oxide	N_2O	Carbon monoxide	CO
Nitrogen oxides	NO_x		

[a] In addition to the nitrogen, oxygen, carbon dioxide, and the inert gases, argon, etc.
[b] The paraffin series given as an example is only one series of a variety of series of hydrocarbons; all of them are composed only of carbon and hydrogen. Many hydrocarbons are released into the air by evergreen trees.
[c] Esters are saltlike products of the reaction between an alcohol and an organic acid.
[d] Ethyl acetate is given as an example; it is the product of the reaction between ethyl alcohol (C_2H_5OH) and acetic acid (CH_3COOH).
[e] R represents a radical or group of atoms, such as CH_3, CH_3CH_2, etc., that maintains its identity in reactions.

Ozone, a natural constituent of tropospheric air, is formed partially during electric storms, partially from reaction between molecular oxygen (O_2) and the atomic oxygen (O) liberated from photodecomposition of nitrogen dioxide present as a pollutant from natural fires. Ozone also diffuses slowly into the troposphere from the stratosphere.

Various organic compounds, listed in Table 9:1, are given off into the troposphere by plants. Some of these undergo photochemical reactions resulting in a number of oxidation products, a few highly noxious

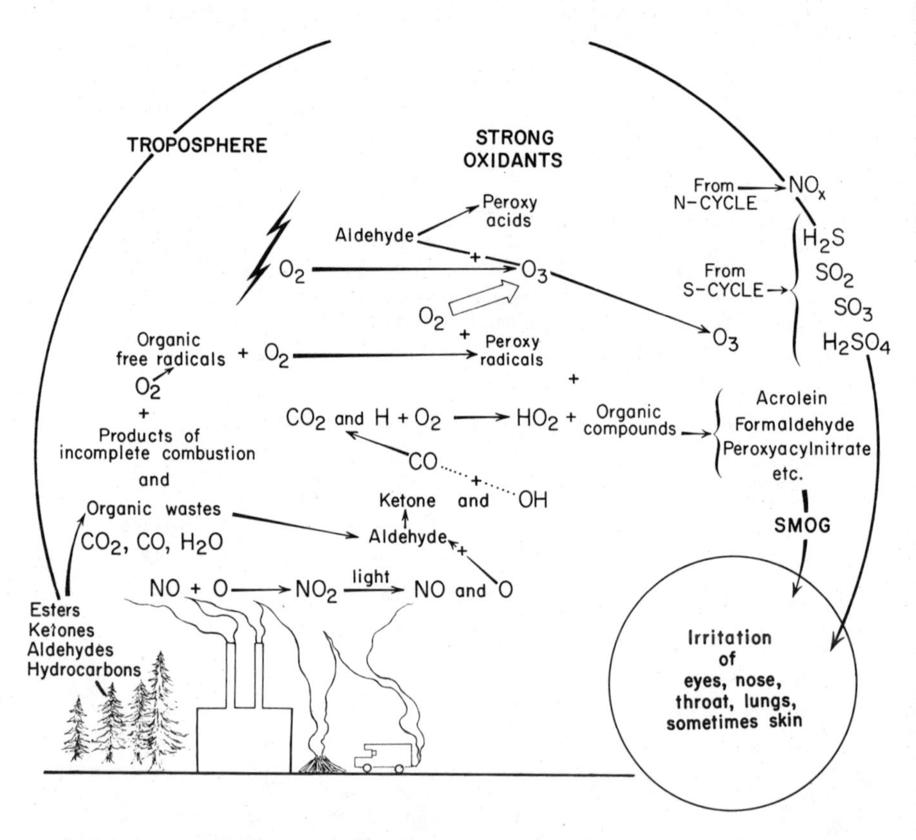

FIGURE 9:5. Some aspects of tropospheric pollution showing formation of various oxidants and their interaction with organic compounds, especially during daylight. In a city, automobiles and factories contribute a major share of pollutants; in an evergreen forest, the trees do. Symbols: NO_X, nitrogen oxides collectively; NO, nitric oxide; NO_2, nitrogen dioxide; O_3, ozone; O_2, molecular oxygen; O, atomic oxygen; CO_2, carbon dioxide; CO, carbon monoxide; H_2O, water; OH, hydroxyl radical; H_2S, hydrogen sulfide; SO_2, sulfur dioxide; SO_3, sulfur trioxide; H_2SO_4, sulfuric acid. Data from Cadle and Allen, *Science* 167 (1970), p. 246.

to life (see Figure 9:5). Before the United States was densely populated and industrialized, a blue haze from release of such organic compounds from trees and forest fires is known to have occurred over mountains—for example, the Smokies and Blue Ridge Mountains—and welled into valleys. Smog generally remains in the atmosphere no more than one or two weeks when atmospheric conditions are favorable. Natural smogs have probably been present since the appearance of heavy land vegetation on the earth's surface, but they did not threaten living things that evolved in their presence. However, smog is exaggerated to the point of intolerance by uncontrolled automobile traffic, industrialization, incineration of wastes from a burgeoning population, and production of synthetic chemicals with which life has had no experience.

Man-Made Tropospheric Smog

In December, 1952, about four thousand people died and thousands of others suffered permanent injuries as a result of one week of exceptionally heavy smog in London. It all started when a dense fog settled over the city and the inversion of warm air over cold prevented vertical movement. Lack of more than slight winds prevented horizontal flushing. Into this deadlocked volume of air the city's furnaces, factories, and automobiles poured tons of smoke, dust, and chemical fumes, turning the fog brown and then black. Visibility fell to a few inches, leading to accidents, and millions went through the city with smarting eyes, coughs, nausea, and vomiting. Some suffered prolonged illnesses. A cleanup of pollutants was demanded and implemented. Today, Londoners live in a much improved atmosphere.

Injury to plants that are sensitive indicators of smog often precedes detectable injury to people. Some plants are injured by one part of hydrocarbon peroxides to 10 million parts of air. They are also injured by as little as one part of sulfur dioxide in 1 million parts of air.

Air pollution also generates fogs. Around large cities fogs have dramatically increased. In Prague, for example, records indicate that fogs have doubled in frequency in eighty years. Thermal pollution, such as runoff of warm waters from industrial cooling systems, nuclear plants, and water cooling towers, are often initiators of fogs. These should therefore be kept at a distance from population centers and airports. Although some scientists claim no relation between fogs and smogs, others believe that they are mutually enhancing. When a fog turns to rain, however, it scrubs the air clean of smog.

Automobiles give off potentially the most offensive compounds in smog. Gasoline consists of a mixture of hydrocarbons from which the energy for propulsion is liberated by combustion. Complete combustion would give rise to carbon dioxide and water, both innocuous compounds, but uncombusted gasoline and products of its incomplete combustion are emitted and nitric oxide also appears. When gasoline contains sulfur, sulfur oxides are present in the exhaust. Industrial and incinerative oxidation of sulfur and nitrogen-containing compounds in organic matter adds noxious sulfur and nitrogen compounds to smog. Though man-made pollutants are not generally different from natural pollutants, they are being added at rates beyond the capacity of winds and rains to flush them from the air we breathe. When pollutants accumulate locally, they become evident by odor, irritation of breathing organs and eyes, reducing visibility, and by injuring and killing plants in their wake. Man-made smog can now be detected in such distant parts of the globe as mid-ocean and arctic and antarctic polar regions.

In daylight most of the products of incomplete hydrocarbon combustion and the nitric oxide in automobile exhaust undergo rapid photochemical alteration in the atmosphere. For example, nitric oxide rapidly converts to nitrogen dioxide (mechanism unknown), which in turn is decomposed by UV-A radiation to nitric oxide and atomic oxygen. Part of the atomic oxygen then reacts with oxygen to form ozone. Some of the atomic oxygen also reacts with hydrocarbons and other organic molecules to produce fragments of organic and inorganic molecules called free radicals. These are unstable and quickly react with radicals or molecules to form a wide variety of secondary products. Among these are peroxy radicals with an extra oxygen and high oxidizing potential, oxidizing nitric oxide to nitrogen dioxide and molecular oxygen to ozone. Most of the ozone in tropospheric smog has this origin (see Figure 9:5). Peroxy radicals also form in other ways. Ozone oxidizes aldehydes to form peroxy acids. Oxygen atoms oxidize aldehydes to form ketones liberating hydroxyl radicals. Hydroxyl radicals react with carbon monoxide to form carbon dioxide and a hydrogen atom (H). Hydrogen atoms react with oxygen to form hydroperoxyl radicals (HO_2). These highly oxidizing free radicals react with organic pollutants in smog to produce a wide variety of irritants. It is likely that formaldehyde (HCHO), peroxyacyl nitrate ($CH_3CO—O—O—NO_2$), and acrolein (CH_2CHCHO) are the main eye irritants in smog.

Figure 9:5 shows the pollutants in industrial air. Their concentrations are given in Table 9:2, and the graph in Figure 9.6 indicates variations of some of them during the day. In each community the magnitude of smog and the relative amounts of various pollutants are related to the kinds and number of industries and the densities of automobile and airplane traffic, as well as the degree of flushing of smog by wind and rain. Scientists have given intensive study to atmospheric pollution in the Los Angeles area. Results of research there and on the U.S. Eastern Seaboard most likely apply in general to most cities, especially when complicated by temperature inversions.

One need not follow the details of chemical reactions in smog to get a feeling for the chemical complexity of the situation; indeed, only a few of the more important reactions are mentioned in this chapter. Keep in mind, however, that every time a new type of molecule is thrust into the atmosphere by human activity, it further complicates the picture. A sum-

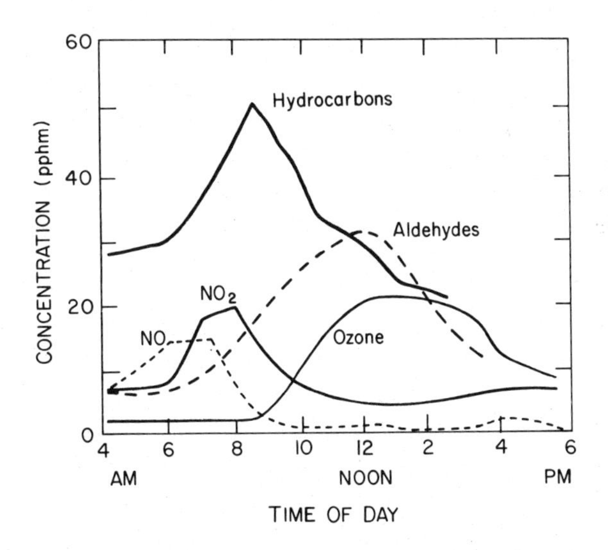

FIGURE 9:6. Typical variation of components in photochemical smog for a day of intense smog; pphm, parts of constituents per hundred million parts of air by volume. Sulfur compounds were not included in the study. For examples of hydrocarbons and aldehydes in such smogs, see Table 9:2. From Leighton, *The Photochemistry of Air Pollution,* Academic Press, New York, 1961.

Table 9:2. Typical Concentrations of Trace Constituents in Tropospheric Smog of an Industrialized Community: Parts of Constituent per Hundred Million Parts of Air by Volume (pphm)[a]

Constituent	Chemical symbol	Concentration (pphm)
Oxides of nitrogen[b]	NO_x	20
Ammonia	NH_3	2
Hydrogen	H_2	50
Water	H_2O	2×10^6
Carbon monoxide	CO	4×10^3
Carbon dioxide	CO_2	4×10^4
Ozone	O_3	50
Hydrocarbons:		
Methane (marsh gas)	CH_4	250
Higher paraffins	C_nH_{2n+2}	25
Ethylene	C_2H_4	50
Higher olefins	C_nH_{2n}	25
Acetylene	C_2H_2	25
Benzene	C_6H_6	10
Aldehydes[c]	RCHO	60
Sulfur dioxide	SO_2	20

[a] After Cadle and Allen, *Science* 167 (1970): 243–249 (table on p. 246).
[b] Oxides of nitrogen, chiefly NO (nitric oxide), NO_2 (nitrogen dioxide), N_2O_5 (nitrogen pentoxide), HNO_3 (nitric acid).
[c] The R in the symbol refers to the rest of the molecule to which the aldehyde group CHO is attached. In acetaldehyde this is CH_3, in propionaldehyde, CH_3CH_2, etc.

mary of the major chemical events occurring in smog is presented in Figure 9:5.

How Tropospheric Smog Reduces Antirachitic UV-B Rays

In California's San Joaquin Valley, where rickets once flourished, blue smog generated by fires and hydrocarbons from large forests in the surrounding mountains combined with man-made smog to diminish the sun's UV-B radiation below the antirachitic dose despite the many sunlit hours. Rickets, you will recall, was also prevalent among children in industrial cities of Europe during the seventeenth and eighteenth centuries; black smoke from industrial chimneys removed the antirachitic radiation from sunlight.

Because the degree and chemical composition of smog vary from time to time and place to place, scientists cannot provide specific data on the degree of absorption of UV-B radiation by tropospheric smog. Furthermore, wind and rain flush smog differently in different topographies and climates. At a given location, scientists simulate the effect of degrees of particulate and chemical smog on the intensity of UV-B radiation by measuring the changes in intensity as the UV-B rays pass through various thicknesses of atmosphere, as happens during the course of the sun's movement from morning to night (see Figure 1:1). Researchers believe this is possible because atmospheric air always contains particulate and molecularly dispersed smog. The greater the depth of atmosphere through which the UV-B radiation must pass, the greater is the degree of absorption, scattering, and reflection of these rays.

How the Nitrogen Cycle Regulates Stratospheric Ozone

Fossil evidence suggests that before an ozone layer was present life was not possible on the earth's surface because the UV-C radiation would have flooded the earth and killed life on land. Only after oxygen was liberated into the atmosphere by photosynthetic plants was invasion of the land by plants and animals possible.

Ozone is synthesized from oxygen in the stratosphere by the absorption of short wavelength UV-C radiation, removing it from sunlight reaching the earth (see Figure 2:2). Molecular oxygen (O_2) in air absorbs UV-C rays (at a peak wavelength of 150 nm) and splits it into two atoms of oxygen. One of these atoms (O) may then combine with another oxygen molecule (O_2) to produce an ozone molecule (O_3). Ozone-forming reactions occur in the presence of a third body, a molecule or a particle. Some ozone is decomposed by reacting with atomic oxygen produced by UV-C splitting molecular oxygen. In this reaction, a molecule of ozone (O_3) and an atom of oxygen (O) react to produce two molecules of oxygen (O_2). Ozone also decomposes by absorption of UV-C radiation of wavelengths longer (peak at 260 nm) than those involved in its synthesis, splitting a molecule of ozone into a molecule and an atom of oxygen. But atomic oxygen almost immediately reacts with molecular oxygen to reform ozone. Thus, much of the UV-C radiation is dissipated as heat with little destruction of ozone in the reaction. In fact, this reaction is largely responsible for heating the ozone layer.

In the laboratory it is possible to determine with pure air the rates of formation and decomposition of ozone in chambers under conditions simulating the atmosphere. From these studies, scientists calculate the equilibrium concentration of ozone. According to these tests, there should be considerably more ozone in the stratospheric ozone layer than is actually found in rocket samples (see Figure 9:7). Something other than the common gases of air must exist in the stratosphere, which decomposes ozone.

In addition to the common gases of air, samples of stratospheric air obtained by aircraft were found to contain nitrogen oxides. When researchers introduced nitrogen oxides in concentrations comparable to those found in the stratosphere, ozone was reduced to approximately the same concentration found in the stratosphere. Clearly, nitrogen oxides must reduce the concentration of ozone from the steady state photochemically possible in pure air to the concentration actually observed in the stratosphere (see Figure 9:7).

How did nitrogen oxides enter the stratosphere? Since no photochemical reaction between the nitrogen and oxygen components of air

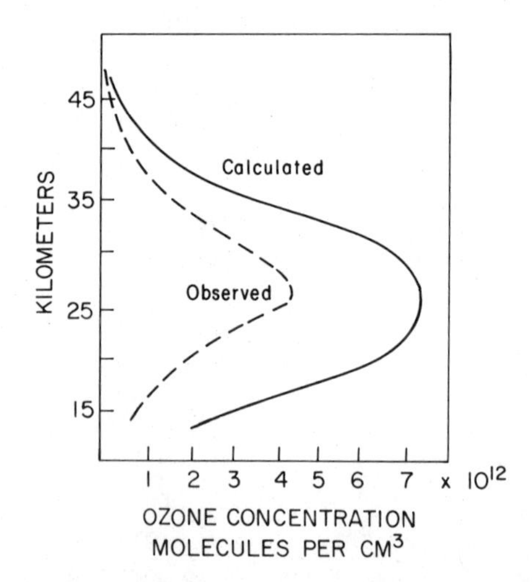

FIGURE 9:7. Calculated ozone profile (solid curve), determined on the basis of photochemical reactions occurring in the stratosphere in the absence of nitrogen oxides (at the equator) and observed ozone profile (dashed curve), resulting from decomposition of some of the ozone by nitrogen oxides. To simplify calculations, a uniform distribution of NO and NO_2 was assumed in the stratosphere as well as a static atmosphere. From Johnston, Proceedings of the National Academy of Science, U.S.A. 69 (1972), p. 2370.

has been found, scientists believe that the nitrogen compound responsible must come from the troposphere. We know that vertical exchange of a turbulent nature between troposphere and stratosphere is minimized by the temperature inversion of the layers, but that slow diffusion exchange between the two occurs continuously. Small amounts of nitrogen oxides are produced by lightning and lesser electrical discharges at the tropopause, but by far the greatest proportion come from nitrous oxide (N_2O) evolved by the continuous operation of the nitrogen cycle in nature (see Figure 9:4). Nitrous oxide also known as the anesthetic, laughing gas, is an intermediate product in the denitrification of nitrates and, while most of it is reduced to molecular nitrogen by denitrifying bacteria, a small amount of it leaks from soil, ocean, and fresh waters rich in organic matter into the air. There it reaches a concentration of 0.25 parts per million. An inert gas with no reaction with air and water, nitrous oxide is unaffected by sunlight in the troposphere. Slow eddies of circulative air diffusion lift it to the tropopause and diffusion takes it into the stratosphere, where it is photochemically decomposed. It can take from ten to seventy years for a nitrous oxide molecule in soil and water to decompose in the stratosphere.

Decomposed photochemically by absorption of UV-C rays high in the stratosphere, nitrous oxide forms molecular nitrogen (N_2) and atomic oxygen (O). Nitrous oxide also reacts chemically with atomic oxygen produced in the photochemical decomposition of ozone in the ozone layer, resulting in a molecule of nitrogen (N_2) and one of oxygen (O_2). However, another equally rapid reaction of nitrous oxide with atomic oxygen causes the formation of two molecules of nitric oxide (NO).

While the quantitative details of the reactions by which nitrous oxide converts to nitric oxide in the stratosphere are not fully elucidated, no one questions the fact that a continuous gradient "pulls" nitrous oxide from the troposphere into the "sink" of the stratosphere. There the nitrous oxide concentration is continuously lowered as it is either destroyed or converted to other compounds, especially nitric oxide.

Nitric oxide formed in this way is extremely reactive and is involved in many reactions in the stratosphere. Of prime interest here is its destruction of ozone (O_3) with which it reacts to form oxygen (O_2) and nitrogen dioxide (NO_2), thereby decreasing the protective ozone layer. We would be less concerned if one molecule of nitric oxide destroyed one molecule of ozone, as would be true if nitrogen dioxide were either inert or removed. Instead, much of the nitrogen dioxide reacts with atomic oxygen to regenerate nitric oxide (see Figure 9:4). Thus, NO returns like a bad penny and cycles many times. Each time it destroys another ozone molecule (such a series of events is known as a chain reaction). Part of the nitrogen dioxide reacts with hydroxyl radicals to form nitric acid, which, on diffusion into the troposphere, is washed out in rains and enters the nitrogen cycle. Figure 9:8 summarizes the reactions that form nitric oxide in the stratosphere and destroy ozone.

While nitrogen oxides other than the few considered above also occur in the stratosphere, details of their interrelationships are of little concern here. Scientists are not certain about all the quantitative aspects, but there is no doubt that nitrogen oxides, chiefly in the form of nitric acid (see Figure 9:4) diffuse from the stratosphere where they are at a

Reactions

$$O_3 + \text{UV-C radiation} \longrightarrow O_2 + O$$

Formation:

$$O + N_2O \longrightarrow NO + NO$$

Destruction of ozone:

$$O_3 + NO \longrightarrow NO_2 + O_2$$

$$O + H_2O \longrightarrow 2\,OH^{\bullet}$$

$$OH^{\bullet} + NO_2 + M \longrightarrow HNO_3 + M$$

$$OH^{\bullet} + HNO_3 \longrightarrow H_2O + NO_3^{\bullet}$$

$$HNO_3 + \text{UV-C radiation} \longrightarrow OH^{\bullet} + NO_2^{\bullet}$$

Reconstitution:

$$NO_3^{\bullet} + \text{red light} \longrightarrow O_2 + NO$$

$$NO_2 + O \longrightarrow O_2 + NO$$

FIGURE 9:8. Formation and reactions of nitric oxide in the stratosphere. Source of nitric oxide, N_2O diffusing from the troposphere. Key to symbols: N_2O, nitrous oxide; O_3, ozone; O_2, molecular oxygen; O, atomic oxygen; NO, nitric oxide; NO_2, nitrogen dioxide; H_2O, water; $OH^{\bullet}$, hydroxyl radical; M, a third body, either a molecule or a particle; HNO_3, nitric acid; $NO_2^{\bullet}$, nitrite radical; H, hydrogen atom, $NO_3^{\bullet}$, nitrate radical. The distinction between singlet atomic oxygen and other atomic oxygen, omitted here for simplicity, as well as other nitrogen oxides and their interactions, are considered in the reference from which the data were taken: Johnston, Proceedings of the National Academy of Science, U.S.A. 69 (1972), p. 2369-2372.

concentration of about 3 to 12 parts per billion to the troposphere where they are at a concentration of 1 to 3 parts per billion. Nitrogen oxides diffuse continuously downward into the troposphere, where they are washed out by rain, creating a further gradient, or sink, for continuation of their downward diffusion (see Figure 9:10).

In a natural state, there is no buildup of nitric oxide in the stratosphere. A steady state is achieved between input of nitrous oxide from the nitrogen cycle, the synthesis of nitric oxides from nitrous oxide in the stratosphere, and the removal of nitric oxides (chiefly as nitric acid), by diffusion into the troposphere. As a result, the natural ozone level in the stratosphere retains a steady state created by the operation of all reactions involved in the synthesis and breakdown of ozone.

In an unpolluted stratosphere, the ozone concentration is regulated by the nitrogen cycle on the surface of the earth. This emphasizes the extraordinarily complex relations that exist in nature and the care that must be exercised before we disturb them. Nitrogen oxides capable of catalyzing the destruction of ozone are lost from the stratosphere primarily by slow diffusion downward, and therefore those injected into the stratosphere by some human activity will remain for appreciable periods—one to three years—depending on the altitude at which they are introduced. Added in this way, nitrogen oxides would accumulate continuously. Of course, some of the nitrogen oxides diffuse upward in the stratosphere and are decomposed by UV-C radiation above twenty-five kilometers. Clearly, unnatural stratospheric pollution by nitrogen oxides, or any other ozone decomposers, is potentially more dangerous than tropospheric pollution because the pollutants are so slowly removed from the stratosphere.

Supersonic Transports

Supersonic transports—SSTs such as the British-French Concorde and the Russian Tupolev—that fly in the stratosphere release nitrogen oxides, particulate sulfates,* uncombusted hydrocarbons, soot, carbon dioxide, carbon monoxide, and water. The amount of water added to the

*Sulfates in SST exhaust form aerosols that, by absorbing light, may modify the temperature of the earth. This effect is not considered in this book. See the National Academy of Sciences report (Booker, 1975) and Bartholic (1975).

stratosphere has been shown to be insignificant in relation to the water vapor already present and is no longer the source of concern that it once was. Of greater concern to us are the nitrogen oxides.

At present, commercial SSTs are essentially experimental. But most government regulatory agencies as well as most aircraft companies believe that a commercial fleet of some five hundred or more supersonic transports carrying long-distance traffic is inevitable. Air lanes in the troposphere are already crowded and few remain to be assigned. Where next?

If air traffic in developed countries continues to increase and an air load is added by developing nations, heavy traffic in the stratosphere is inevitable. The only way in which future air traffic can increase without a corresponding increase in airplane accidents is by opening more air lanes; this means mainly higher lanes, lanes for SSTs in the stratosphere. There is even talk of hypersonic planes that will fly at one hundred thousand feet, well into the stratosphere and at about the peak of the ozone layer. The higher the flight altitude, the longer it will take to remove the exhaust products ("air garbage," if you will) from the atmosphere. Unless pollutant levels present in the exhaust of modern planes are lowered by improved engine design, the hazard to the ozone layer from future air traffic will be far greater than that from the present air traffic. We are now only slowly invading the lower levels of the stratosphere with the high-flying subsonic planes (the 747, for example, on part of a transatlantic flight) and the small number of military and civilian supersonic planes currently in service. It is, therefore, of great importance to assess the possible effects of an SST fleet on the stratosphere and upon the earth as a whole before it is allowed to happen. A number of estimates on the possible pollution of the stratosphere and discussions of possible effects on bacteria, plants, animals, and man have already been made. The National Research Council, The National Academy of Sciences, the National Academy of Engineering, the Department of Transportation, and the Institute of Defense Analyses in the United States* and similar agencies abroad have taken the lead in analyzing the available facts.

The quantity of pollutants in the exhaust of present supersonic transports leads scientists to estimate that the amount of nitrogenous

*The Department of Transportation has terminated its program of studies on atmospheric pollution but the National Aeronautics and Space Administration and the National Oceanic and Atmospheric Administration are continuing studies on some aspects.

Table 9:3. Quantity of Ozone-Destroying Pollutants Calculated to Reach the Stratosphere Annually[a]

Substance	Source	Quantity (megatons)
NO_x[b]	Nitrogen cycle in nature	1.0
NO_x	500 present-type Concorde or Tupolev SST planes	0.5
NO_x	500 1971-type Boeing SST planes	1.8
ClO_x[c]	Freon propellants CCl_2F_2	0.5 (1974)
ClO_x	50 space shuttle flights	0.003

[a]Data from Harold Johnston, 1975. Personal communication.
[b]Chiefly NO, NO_2 and HNO_3.
[c]Chiefly Cl, ClO and HCl. The figure given is for 1974. A larger amount will be liberated in successive years unless controls are imposed.

pollutants liberated into the stratosphere from a commercial fleet of five hundred supersonic transports of the British-French Concorde and Russian Tupolev type would reduce the ozone layer by about 3 to 4 percent. A projected American-designed supersonic fleet of five hundred planes with greater fuel consumption would have reduced the ozone layer by 17 to 18 percent (Table 9:3). The present subsonic plane incursions into the lower stratosphere have a negligible effect on the ozone layer. However, jet engine evolution has been toward higher combustion temperatures, which means a larger amount of nitrogen oxides produced per unit of fuel consumed.

If a fleet of commercial supersonic airplanes is inevitable, it is necessary to lower the combustion temperature in the engine to reduce the combination of nitrogen and oxygen of the air, the main source of nitrogen oxides in the exhaust. This is considered feasible but expensive because it requires redesigning the engines. But there is only one protective ozone layer. Consider it priceless.

Atomic Bombs

At the high temperature generated by an atomic bomb explosion, oxygen and nitrogen in the air combine to form nitrogen oxides. In addition, bombardment by emanations from residual radioactivity in the air induces further formation of nitrogen oxides. Even atomic bomb explosions at sea level inject some of the nitrogen oxides into the stratosphere. When an atomic bomb explodes high in the atmosphere, perhaps even in the stratosphere, removal of the nitrogen oxides (and other wastes) from the stratosphere to the troposphere where washout may occur is very slow. Some evidence indicates that the tests by the United States and the Soviet Union in the early sixties (1961–1962) reduced the stratospheric ozone concentration for several years (Figure 1:4). There is now concern that further atomic bomb testing will add an undesirable load of nitrogen oxides to that already contaminating the stratosphere from high-flying planes. It is more and more evident that in an atomic war there would be no victors, since in addition to the combatant casualties at the site of the explosion, there would be worldwide noncombatant casualties resulting from impairment of the ozone layer. A greatly reduced ozone concentration would permit more intense UV-B and probably UV-C radiation to penetrate to sea level, a distinct threat to all life on earth.

The report, "Long-term Worldwide Effects of Multiple Nuclear-weapons Detonations," issued by the National Academy of Sciences in late 1975 estimates that if roughly half the atomic arsenal of the two major atomic powers were released the 50 to 75% decrease in the ozone layer would have even more disastrous global effects than the ionizing radiations which would be devastating primarily in the locality of the strikes.

Spray Can and Refrigerator Freons

Recently another surprising source of pollution has been shown to destroy ozone. It is atomic chlorine liberated from Freons (chlorofluoromethanes), synthetic compounds not known to occur in nature, used as propellants in spray cans and as refrigerants. Freons are synthesized from methane, also known as marsh gas (CH_4), in which chlorine and fluorine atoms substitute for hydrogen atoms, the two chief forms being dichlorodifluoromethane (CF_2Cl_2) and trichlorofluoromethane ($CFCl_3$). Freons are inert and are therefore neither decomposed nor removed in the troposphere. Thus, they have plenty of time to rise to the stratosphere. There, Freons, which strongly absorb in the region of 220 nm in a "window" (see Figure 2:2) between the strong ab-

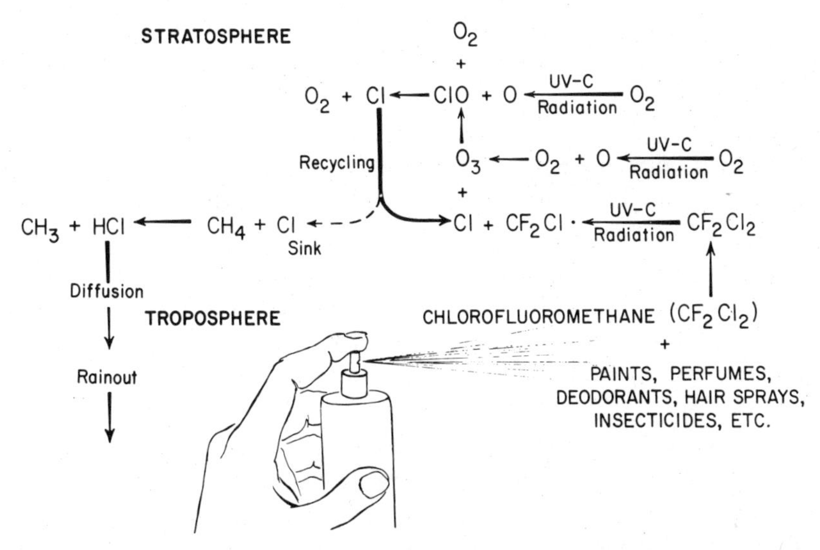

FIGURE 9:9. Diagram of the presumed effect of Freons (chlorofluoromethanes) used as propellants in spray cans and refrigerators on the decomposition of ozone in the stratosphere.

sorption bands of oxygen and ozone, are decomposed by UV-C rays, liberating chlorine atoms (Cl) and CX_3 (where X stands for either chlorine or fluorine). Chlorine atoms react with ozone to form oxygen (O_2) and chlorine oxide (ClO). The latter compound reacts with atomic oxygen (a product of the photodecomposition of oxygen or ozone molecules by UV-C rays), forming oxygen molecules and, again, chlorine atoms. Because chlorine atoms, like nitric oxide molecules, are regenerated for another cycle of ozone breakdown, the process is self-sustaining (catalytic) in a chain reaction (see Figure 9:9). Fluorine is liberated in similar reactions; it also reacts with ozone but fails to produce chain reactions.

Because of their increasing use, presumably without removal from air except for the undetermined amount of chlorine removed as hydrochloric acid (HCl), the concentration of Freons is known to be progressively rising in the troposphere, and therefore also probably in the stratosphere. In fact, the impact of the present tropospheric load of Freons has not yet been felt because of its slow upward diffusion through the tropopause. The prediction of atmospheric scientists, based on the present rate of increase in manufacture and liberation of Freons into the troposphere, is that in a few years the rate of decomposition of ozone may equal the rate of breakdown from operation of the natural nitrogen cycle (N_2O) entry into the stratosphere. Despite the predicted danger of these events to all of us, to our crops on land and sea, no steps have yet been taken to control Freon manufacture and use. Regulatory agencies and legislators sidestep the issue by claiming that no adequate measurements on the rate of Freon increase in samples of stratospheric air and its effects on the decomposition of ozone there have been made. A committee to study the problem was appointed by the National Academy of Sciences in November, 1974.* Some atmospheric scientists believe that in all probability Freons will decompose ozone as effectively in the stratosphere as in the laboratory. It is, of course, possible that Freons are eliminated in some still undiscovered atmospheric reactions other than the stratospheric photodecomposition described, but the stakes are too high to be comfortable with this hope. They are not removed from the troposphere by prolonged contact with water, soil, or plants, though their concentration is said to decrease somewhat after reaction with heavy smog.

For industrial purposes, chlorine is introduced into a large number of organic compounds and many of these compounds are highly volatile—for example, carbon tetrachloride (CCl_4) and chloroform ($CHCl_3$)—and enter the atmosphere. It has been pointed out that these and other chlorinated compounds have been used for a long time in increasing amounts without any known effect on the ozone layer. However, some of these compounds react with chemicals in the atmosphere, while others are taken up readily by cells of organisms, as in the case of carbon tetrachloride, which is highly poisonous, and chloroform, which serves as an anesthetic. Such reactive compounds are probably removed from the atmosphere and do not accumulate like the inert Freons. Nevertheless, tests for their presence in the stratosphere are desirable as they are also possible sources of chlorine atoms. Volatile brominated compounds, widely used in industry, should also be scrutinized for possible action on ozone. Methyl chloride (CH_3Cl) formed by fermentative microbes may serve as another source of chlorine atoms polluting the stratosphere, although little information is available at present.

Chlorine atoms are liberated from ammonium perchlorate used as an oxidizing agent in rocket engines of the space shuttle. While at present

*A summary of the committee's findings will appear in a "Report of the Federal Task Force of Inadvertent Modification of the Stratosphere," to be issued by the Council on Environmental Quality in early 1976. Some industrial firms are sponsoring research projects aimed at finding possible sinks for Freons in both troposphere and stratosphere.

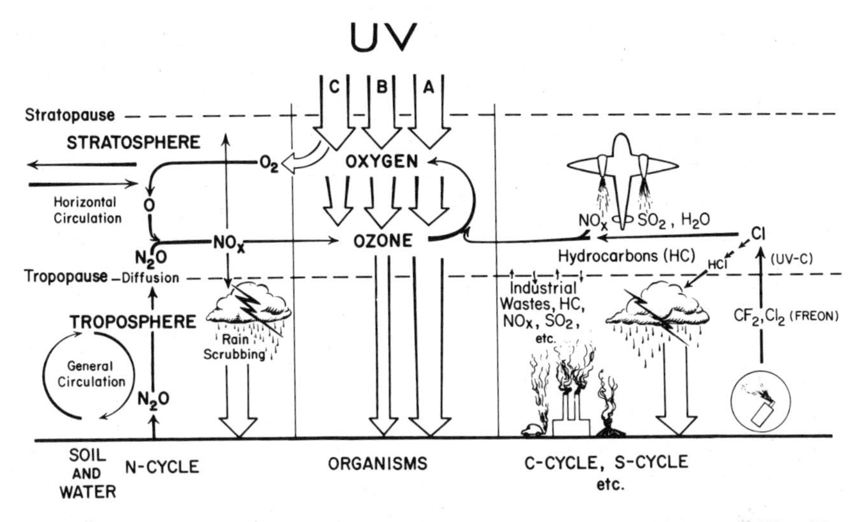

FIGURE 9:10. Diagram of the probable interrelations of the earth's surface and life with the atmosphere and sunlight.

injection of chlorine into the stratosphere from this source is small, it is not negligible and should also be taken into account. Too little is yet known of its consequences to impose controls.

What Are the Consequences?

We now know that ozone shields us from the sun's deadly ultraviolet radiation as it strikes the outer surface of our atmosphere (see Figure 9:10). Reducing stratospheric ozone, even modestly, poses a potential threat to all life—human, plant, animal, and bacteria. Slight decreases in the ozone layer effectively increase UV-B ray intensity reaching the earth. A large ozone decrease* would not only permit more UV-B radiation to penetrate, but also permit some of the sun's UV-C rays to reach the earth, rays that at present do not reach sea level (except possibly during some extreme solar flares). Atmospherically, ozone reduction would turn the clock of geological time backward to an era when the ozone layer covering the earth was much thinner. Total destruction of the ozone layer would be inconsistent with life on the earth's surface (see Figure 2:4).

*Even if the ozone shield were attenuated, UV-A and visible radiation would remain much the same as at present. Neither is markedly absorbed by ozone. Infrared radiation, which is absorbed by ozone, would increase somewhat.

While one cannot imagine that a drastic reduction of the ozone layer (or its elimination) would be permitted, smaller reductions in the ozone layer as a consequence of stratospheric pollution, however, might occur before action is taken.

What would happen? The remote possibility exists that earth life will eventually evolve to produce organisms more resistant to damage from increased UV-B radiation. But we cannot even make educated

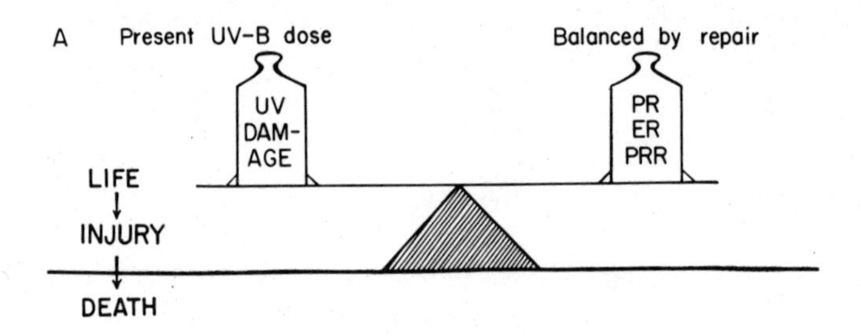

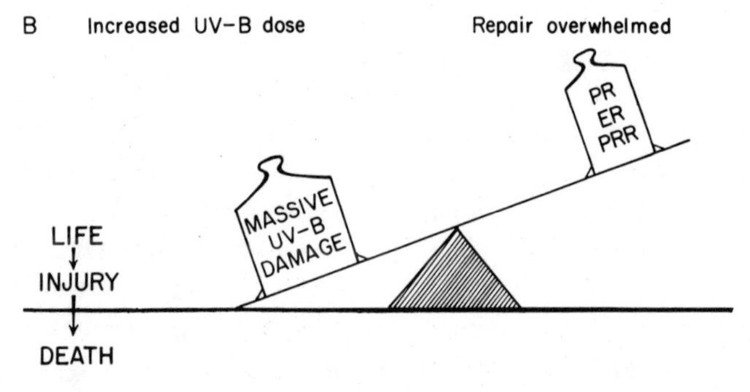

FIGURE 9:11. Life represented schematically as existing in an uneasy balance between damage and repair. At the present UV-B radiation level, photoactivation repair (PR), excision repair (ER), and postreplication repair (PPR) keep cells healthy, counteracting continuously occurring damage from sunlight. If the UV-B radiation were to increase by a small degree, as would follow a small decrease in the ozone layer, repair would be incomplete and cell functions would decline, as indicated by the arrow into the "damage" zone. If the UV-B radiation were to increase greatly as a result of a greater decrease in the ozone layer, the repair systems could be overwhelmed and death would result, assuming no mutants with better repair systems were to appear. (UV-A radiation would not increase appreciably with a decrease in the ozone layer, and its effect is therefore not considered.)

guesses about such a turn of events because no one knows enough yet about the long-term effects of these rays. The damage caused by SSTs, Freons, and nuclear bombs would probably be so rapid that even microorganisms would not have a chance to evolve.

As Chapter 4 revealed, UV-B radiation destroys all types of cells, including bacteria, transparent plant and animal cells, and cells in the unpigmented skin of humans and animals. Chapters 5 and 6 disclosed that should we increase the intensity of UV-B radiation from pollution of the stratosphere, we would accelerate sunburn and chronic skin damage. And, as Chapter 8 indicated, skin cancers—the most common human cancers—would increase proportionally as UV-B rays increase. Dermatologists expect a 2 percent increase in cancer for each 1 percent decrease in ozone. No one yet knows what effect all this would have on our forests, crops, orchards, and pollinating insects. Nor do we know how rapidly our natural and synthetic polymers would deteriorate. Neither do we know how sensitive microscopic organisms (plankton) and other sea- and freshwater life is to such radiation, and so we cannot predict the effect of an increase in UV-B radiation on the aquatic crop that feeds our valuable shellfish and fish supply. And, finally, we do not know to what extent entire assemblages of organisms, called ecosystems, and the relations of ecosystems to one another would be affected by these radiation changes.

Scientists do know, however, as Chapter 4 outlined, that life is held in balance with the current UV-B radiation dose level largely because of repair mechanisms. An overdose of UV-B radiation could overwhelm the repair mechanisms, resulting in injury or death to many forms of life (see Figure 9:11). In Chapter 4 we discovered that mutations, generally adverse, could be greatly increased in many lower organisms.

If we permit pollution of the stratosphere by uninhibited flying of unmodified SST fleets, if we continue to escalate the use of Freons, if we permit continued atomic bomb tests and unlimited space shuttles in defiance of the threat to the ozone layer, the effects could be catastrophic. Even if we could reverse these trends, nonetheless at the least they would be hazardous and unpredictable. Undoubtedly, atmospheric changes could add dramatically to society's burden, already intense because of overpopulation, resource depletion, economic and political instability.

What is needed at the earliest possible moment is international cooperation to halt atmospheric pollution and to preserve a reasonably clean stratosphere. Certainly, the problem is global. Pollution of the stratosphere is not restricted to local areas as is tropospheric contamination although at present it is primarily of concern to the northern hemisphere since north–south movement in the stratosphere is slow. Industry, government, and the scientific community should immediately come to grips with this threat to our survival before we seriously damage or destroy our atmospheric shield. How many spray cans, SST flights, or atomic explosions will it take to leave us unprotected from death by ultraviolet rays?

Our future lies vulnerably in our hands. If we act now to preserve the stratosphere, and thus to protect life on earth from increased exposure to ultraviolet radiation, then we will have learned to live with our sun's ultraviolet rays.

For Additional Reading

Bartholic, J., ed. "Impacts of Climatic Change on the Biosphere. CIAP Monograph 5. Part 2. Climatic Effects." Department of Transportation. Climatic Impact Assessment Program. Washington, D.C., 1975.

Booker, H. C., ed. *Environmental Impact of Stratospheric Flight. Biological and Climatic Effects of Aircraft Emissions in the Stratosphere.* Washington, D.C.: National Academy of Sciences, 1975.

Broderick, A. J., and Hard, T. M., eds. *Proceedings of the Third Conference on the Climatic Impact Assessment Program,* Washington, D.C.: U.S. Dept. of Transportation, 1974.

*Cadle, R. D., and Allen, E. R. "Atmospheric Photochemistry." *Science* 167 (1970): 243–248.

*Cicerone, R. J., Stolarski, R. S., and Walters, S. "Stratospheric Ozone Destruction by Man-Made Chlorofluoromethanes." *Science* 185 (1974): 1165–1166.

Cutchis, P. "Stratospheric Ozone Depletion and Solar Ultraviolet Radiation on Earth." *Science* 184 (1974): 13–19.

Delwiche, C. C. "The Nitrogen Cycle." *Scientific American* 223 (Sept., 1970): 137–146.

Dutsch, H. U. "Photochemistry of Atmospheric Ozone." *Advanced Geophysics* 15 (1971): 219–322.

Giese, A. C., and Christensen, E. "Effects of Ozone on Organisms." *Physiological Zoology* 27 (1954): 101–115.

Goldsmith, J. R. "Effects of Air Pollution on Human Health." In *Air Pollution* vol. 2, 2nd ed., edited by A. C. Stern. New York: Academic Press, 1968, pp. 547–615.

Hammond, A. L., and Maugh, T. H. II. "Stratospheric Pollution: Multiple Threats to Earth's Ozone." *Science* 186 (1974): 335–338.

Hampson, J. "Photochemical War on the Atmosphere." *Nature* 250 (1974): 181–191.

Handler, P. ed. "Long-term Worldwide Effects of Multiple Nuclear-weapons Detonations." National Academy of Sciences, Washington, D.C., 1975.

Hobbs, P. V., Harrison, H., and Robinson, E. "Atmospheric Effects of Pollutants." *Science* 183 (1974): 909–915.

Johnson, F. S. "SST, Ozone and Skin Cancer." *Astronautics and Aeronautics,* July, 1973: 16–21.

Johnston, H. "Reduction of Stratospheric Ozone by Nitrogen Oxide Catalysts from Supersonic Transport Exhaust." *Science* 173 (1971): 517–522.

———. "Newly Recognized Vital Nitrogen Cycle." Proceedings of the National Academy of Science, U.S.A. 69 (1972): 2369–2372.

Leighton, P. A. *Photochemistry of Air Pollution.* New York: Academic Press, 1961.

McConnell, G., Ferguson, D. M., and Pearson, C. R. "Chlorinated Hydrocarbons and the Environment." *Endeavour* 34 (1975): 13–18.

*Molina, M. J., and Rowland, F. S., "Stratospheric Sink for Chlorofluoromethanes. Chlorine Atom-Catalyzed Destruction of Ozone." *Nature* 249 (1961): 810–812.

Myers, J. N. "Fog." *Scientific American* 219 (Dec., 1968): 75–82.

Nachtwey, D.S., ed. "Impacts of Climatic Change on the Biosphere. CIAP Monograph 5. Part 1. Ultraviolet Effects." Department of Transportation. Climatic Impact Assessment Program. Washington, D.C., 1975.

Rosenthal, H. *Selected Bibliography on Ozone, Its Biological Effects and Technical Applications.* Fisheries Research Board of Canada, Technical Report #456, 1974.

*Smith, K. C. *Biological Impacts of Increased Intensities of Solar Ultraviolet Radiation.* Washington, D.C.: National Academy of Sciences, and National Academy of Engineering, 1973.

Stockinger, H. E., and Coffin, D. E. "Biologic Effects of Air Pollutants." In *Air Pollution,* vol. 1. 2nd ed., edited by A. C. Stern. New York: Academic Press, 1968, pp. 445–546.

Went, F. W. "Air Pollution." *Scientific American* 192 (May, 1955): 63–72.

Wilkness, P. E., Swinnerton, J. W., Lamontagne, R. A., and Brissan, D. J. "Trichlorofluoromethane in the Troposphere, Distribution and Increase, 1971 to 1974." *Science* 187 (1975): 832–834.

Woofsy, S. C., McElroy, M. B., and Sze, N. D. "Freon Consumption: Implications for Atmospheric Ozone. *Science* 187 (1975): 535–537.

Index

An italicized number refers to a figure. Letter T to the right of a number refers to a table.